FOUNDATION ELECTRONICS

Nelson

Author's Acknowledgements

The author wishes to acknowledge the assistance given by the following, at various stages of the book's development:

A. F. Abbott, for the encouragement to make a start;

A. Balatoni, for constructive criticism in the early stages;

Mrs K. Lawson, for much of the typing;

Bob Spence of Imperial College, for checking the semiconductor sections;

Dr. J. Calvert of Manchester University for expert guidance in the final stages of preparation.

N. Cull of de Stafford School, for proof reading.

Thomas Nelson and Sons Ltd
Nelson House Mayfield Road
Walton-on-Thames Surrey
KT12 5PL UK

51 York Place
Edinburgh
EH1 3JD UK

Thomas Nelson (Hong Kong) Ltd
Toppan Building 10/F
22A Westlands Road
Quarry Bay Hong Kong

Distributed in Australia by
Thomas Nelson Australia
480 La Trobe Street
Melbourne Victoria 3000
and in Sydney, Brisbane, Adelaide and Perth

First published by Thomas Nelson and Sons Ltd 1981
ISBN 0-17-448121-7
NPN 9 8 7 6 5 4

Printed in Hong Kong

Preface

This course of electricity and electronics is suitable for those students who, having completed O-level or CSE courses in physics, wish to study electronics in a little more detail. It is designed for students aged 16 + who are following a one year course. It is an ideal alternative O-level examination course for A-level physics students to follow during their first year, there being sufficient overlap of content for this not to require too much additional work or apparatus. As the course combines electricity with electronics, it provides a more useful foundation than a pure electronics course.

For students who are thinking about starting this course the only essential requirement is some previous interest or experience in electronics. No knowledge of physics is assumed and very few mathematical demands are made beyond routine calculations. Only limited skill will be necessary to complete the practical project demanded by the examination syllabus.

This book covers the topics in a thoroughly practical way, but experimental detail has been kept to a minimum so that the main narrative is not broken up. The approach is direct and visual, there being no faster way to the understanding of concepts than to have the right images in mind. British Standard symbols have been used throughout, with the exception of the logic symbols, where the more popular American symbols have been adopted with slight modifications, and the crossing of conductors, where in ambiguous cases dots indicate conduction and gaps indicate non-conduction, as the British Standard can cause confusion.

The units are essentially self-contained and they can be studied in almost any order. However, units one to five must be regarded as essential core material, to be thoroughly understood before attempting the remainder.

B. G. Barker
December 1979

Contents

Charge current and potential difference

Unit 1

Introduction

The effects of static electricity have been known since at least 600 B.C. when the Greek philosopher, Thales of Miletus, showed that amber rubbed with fur would attract feathers; he also knew of lodestone, the naturally magnetised iron ore used by the Phoenician mariners as a magnetic compass.

It was not until A.D. 1600 that Dr Gilbert, Queen Elizabeth's physician, first used the word electricity to distinguish the forces between magnets and between static electric charges. In 1800, Galvani and Volta showed that charges would flow in a metal wire, and in 1820 Oersted showed that an electric current could produce magnetic effects. In 1879, Thomas Edison made his first electric lamp and in 1883, whilst trying to improve the performance of his lamps, he accidentally discovered that electricity could flow through a vacuum. This was the beginning of electronics, which is the study of electron flow in materials other than metallic conductors.

The structure of matter

Since the time of the early Greek civilisation, people have believed matter to be made from minute **atoms**. More recently we have discovered not only that atoms exist, but that they in turn are built up from even smaller particles called **protons**, **neutrons** and **electrons**. Modern research has shown these sub-atomic particles to be built from even smaller particles, but it is still possible to study the electrical nature of matter in terms of the proton, the neutron and the electron.

Electric charges

The smallest charge which can exist is the electron and it is written $-e$. The smallest positive charge is carried by a proton and since its value is equal and opposite to that of an electron, it is written $+e$. The neutron carries no electric charge.

The atom as a whole is electrically neutral, so it must always have an equal number of protons and electrons. Thus, under most conditions, ordinary objects around us are electrically neutral. *Figure 1.1* shows, in simplified form, how the sub-atomic particles are arranged in the atom.

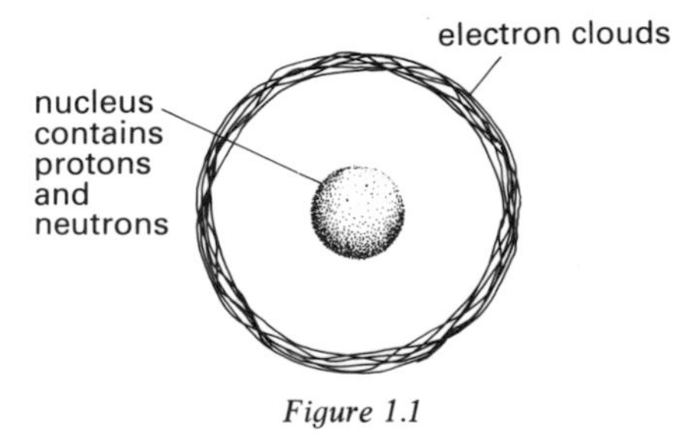

Figure 1.1

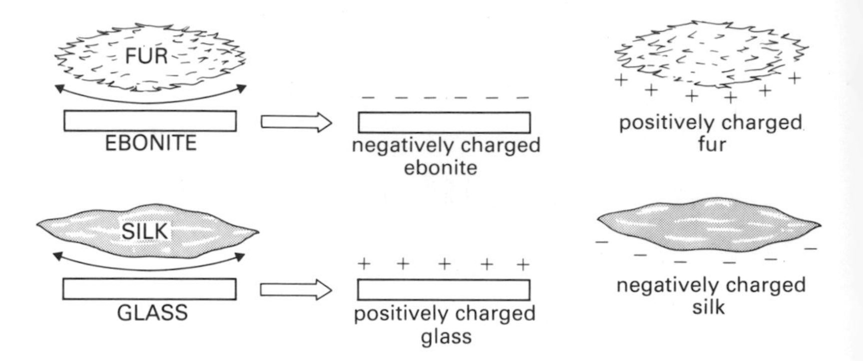

Figure 1.2a Producing a negative charge on ebonite. In practice, huge numbers of electrons are involved in static charges.

Figure 1.2b Producing a positive charge on glass

If two objects are rubbed together, electrons may be transferred from one to the other, as shown in *figure 1.2*. In these diagrams each positive (+) charge is best considered as a hole created by the absence of an electron. If one electric charge approaches close to another, there will be a force of either attraction or repulsion. The rule is:

like charges repel opposite charges attract

An atom which has gained or lost electrons is no longer electrically neutral; it is then known as an **ion**. If an atom loses electrons it becomes a positive ion and if it gains electrons it becomes a negative ion.

The electric current

When we use the word electricity we usually mean current electricity, which is a flow of electric charge. Substances through which an electric current can flow easily are called **conductors**. An electric current is a **drift of electric charge** produced when a conductor is connected across the terminals of, for example, a battery. There are, however, some important differences between the currents in solids, liquids and gases, which are illustrated in *figures 1.3–5*.

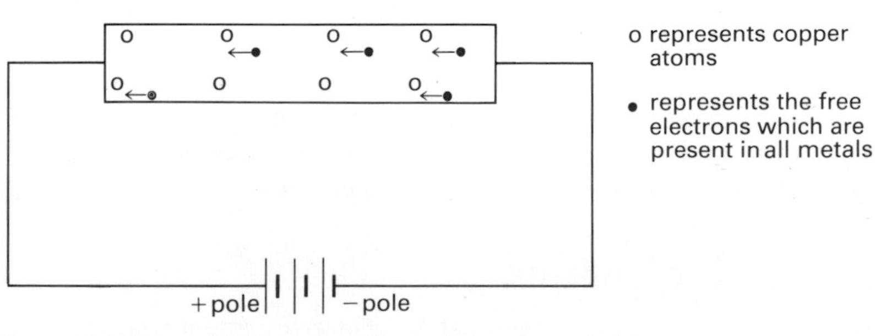

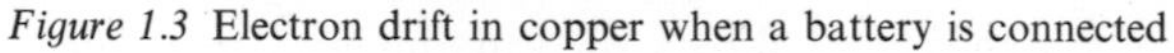
Figure 1.3 Electron drift in copper when a battery is connected

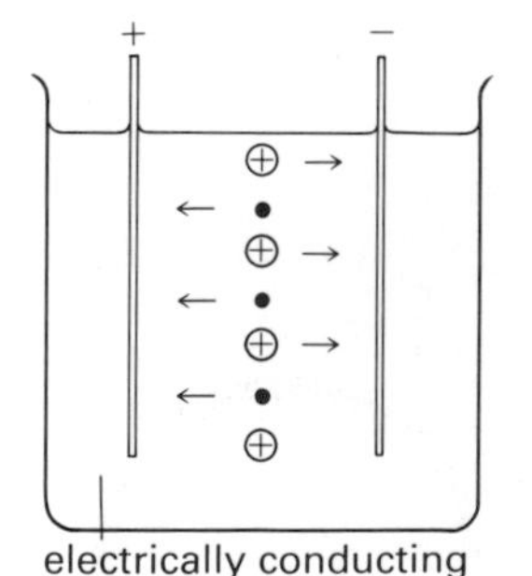

Figure 1.4 Movement of charges in an electrolyte

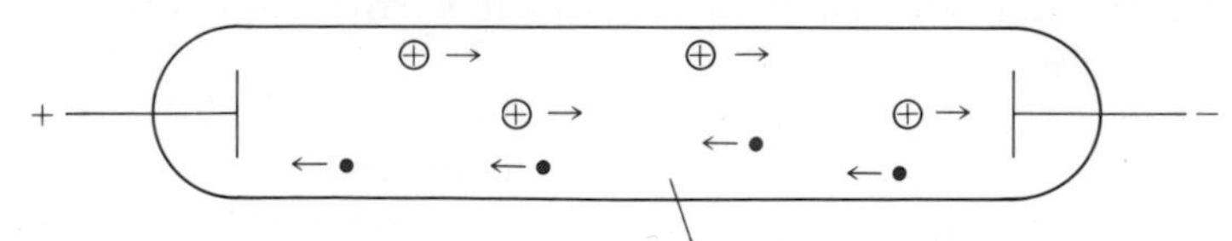

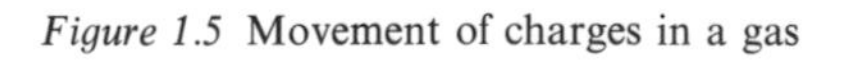
Figure 1.5 Movement of charges in a gas

Notice that in *figure 1.3*:

There are no free positive ions.
The copper atoms do not move.
The electron drift is from the negative (−) pole to the positive (+) pole, outside the battery.

Compare this with *figures 1.4–5* in which

there are both positive and negative ions.
the ions move in opposite directions under the influence of the battery.

It should also be noted that electric charges can flow across a vacuum, as demonstrated by a television tube.

Conventional current

When scientists first began working with electric currents, they believed something was flowing, but had no way of knowing what it was or which way it was moving. They assumed that an electric current in a wire was a flow of positive charge from the positive side to the negative side of the battery connected to it. We now call this flow the **conventional current**, and we continue to make use of the idea, because of the many rules drawn up in the past which depend upon it.

The units of current and charge

The unit of current is the **ampere** (A); it is defined in terms of the force of attraction between two wires, each carrying 1 ampere, thus:

The ampere is the steady current which, flowing in two very long straight, parallel wires placed 1 metre apart in a vacuum, produces a force of 2×10^{-7} newtons per metre length of the wire.

Do not worry too much about the complexities of this definition. At the National Physical Laboratory there is a **current balance** which measures a current by means of the force it produces when flowing in coils of wire. We measure current in the laboratory by using an **ammeter**, which also relies on this force. When small currents are being measured, the units **milliampere** (mA) and **microampere** (μA) are used.

$$1\,\text{mA} = 0{\cdot}001\,\text{A}$$

$$1\,\mu\text{A} = 0{\cdot}001\,\text{mA} = 0{\cdot}000\,001\,\text{A}$$

notice also that 0·001 A may be written 10^{-3}A and 0·000 001 A may be written 10^{-6}A.

The unit of charge is the **coulomb** (C). It represents the quantity of electricity which passes a point in a circuit when a steady current of 1 ampere flows for 1 second.

1 ampere = 1 coulomb/second

It is interesting to note that it requires about 6×10^{18} electrons to carry a charge of 1 coulomb. Even more surprising, perhaps, is the very low drift velocity of electrons in a wire. For example, if a copper wire of diameter 0.5 mm was carrying a current of 1 A, the electrons would, on average, be moving through the wire at less than 0.5 mm per second. This is a tremendous contrast to the velocity of electrons striking the screen of a television tube, where they have reached about one quarter of the speed of light. (Light travels 300 000 000 metres in 1 second.)

Potential difference (p.d.)

An electric charge will only flow between two points if there is a difference of electrical pressure, or **potential difference**, between them. There are useful similarities here with water flow and heat flow; water requires a pressure difference to make it flow and heat energy will only flow if there is a temperature difference. In order to move an electric charge it must be pushed, and if movement does occur we say **work** is done. Work is always done if a force produces movement. Work is measured in a unit called the **joule**.

Two points in a circuit have a potential difference (p.d.) of 1 volt if 1 joule of work is done in moving 1 coulomb of charge between those points.

For small values of p.d. the unit is the **millivolt** (mV) and for large values the **kilovolt** (kV)

1 mV = 0·001 V

1 kV = 1000 V

Electromotive force (e.m.f.)

Every source of electrical energy is really a **transducer** which acts by converting one form of energy into electrical energy.
A battery converts chemical energy into electrical energy.
A dynamo converts mechanical energy into electrical energy.
A thermocouple converts heat energy into electrical energy.
The e.m.f. of these devices is a measure of their ability to convert energy.

The e.m.f. of a source is the total work done in joules if 1 coulomb of charge is moved round the circuit, including through the source.

When a battery, for example, drives a current round a circuit, some of its energy is wasted as it drives that current through itself. This energy appears as heat in the battery and not as useful electrical energy. The energy that is left to drive 1 coulomb through the rest of the circuit is called the **terminal p.d.** of the battery. The terminal p.d. will be less than the e.m.f. when current is being driven and equal to the e.m.f. if the battery is driving no current. This is true for all sources of e.m.f.

It follows from this last statement that the only types of voltmeter which can measure e.m.f. are those which draw no current from the circuit in which they are connected. When using a voltmeter to test a battery, the battery should be made to drive a reasonable current by connecting a resistor across it. A resistor of value about 100 Ω should be connected across a 1.5 V cell before measuring the terminal p.d. with a voltmeter. This 'on load' test will give a much more reliable indication of the condition of the cell than simply connecting a voltmeter across its terminals.

Direct and alternating current

A **direct current** (d.c.) is one which flows in one direction only. A battery drives a direct current because it can only push in one direction. The resulting (conventional) current flows from the positive terminal, and then round the circuit to the negative terminal. The current may vary in strength, but provided it is always flowing in one direction, it is d.c.

A device such as a cycle dynamo or a car alternator produces an **alternating current** (a.c.). In this case the current reverses its direction of flow at regular intervals. If the direction changes less than about 100 times every second, the a.c. would be described as having a **low frequency**. A **high frequency** a.c. may change direction 1 000 000 times every second, or even more.

Unit 2

Resistance

In this unit, you will meet for the first time the two word endings **-or** and **-ance**. A **resistor** is a device which has the special property called electrical **resistance**. The same two endings will be seen in later units on capacitors and inductors.

The use of the word resistance immediately suggests some form of opposition. The resistance of any conductor is the obstruction offered by this conductor to the flow of current.

Units of resistance and Ohm's law

The unit of resistance measurement is the **ohm** (Ω). For larger values of resistance the kilohm (kΩ) and the megohm (MΩ) are used.

$$1\,\text{k}\Omega = 1000\,\Omega$$

$$1\,\text{M}\Omega = 1000\,\text{k}\Omega$$

$$\begin{array}{c}\text{resistance of a}\\ \text{conductor in ohms}\end{array} = \frac{\text{p.d. across it in volts}}{\text{current through it in amps}}$$

By using letters, we can write this equation

$$R = \frac{V}{I}$$

but it is probably easier to remember it in the form

$$V = IR$$

Notice the use of the letter I to represent an unknown current and take care not to confuse it with the number 1. With the aid of this equation, you can calculate the value of any one of the quantities, provided you know the values of the other two. Take care that the units you use are always:

volts, **ampères** and **ohms**

Example To calculate the current driven by a p.d. of 5 mV through a 1 kΩ resistor:

$$V = 5\,\text{mV} \qquad\qquad R = 1\,\text{k}\Omega$$
$$= 0{\cdot}005\,\text{V} \qquad\qquad = 1000\,\Omega$$

$$V = IR$$
$$0{\cdot}005 = I \times 1000$$
$$I = \frac{0{\cdot}005}{1000}$$
$$= 0{\cdot}000\,005\,\text{A}$$
$$\therefore\ I = 5\,\mu\text{A}$$

Ohmic and non-ohmic resistors

A linear, or ohmic resistor, is one which obeys **Ohm's law**, that is, one for which the ratio V/I is a constant, provided the temperature does not change. A non-linear, or non-ohmic resistor, does not obey Ohm's law. The graphs in *figure 2.1* illustrate this.

Metallic conductors and carbon obey Ohm's law if their temperature is kept constant. Lamp filaments, cells, electrolytes and a special class of substances called semiconductors (see unit twelve), do not obey Ohm's law.

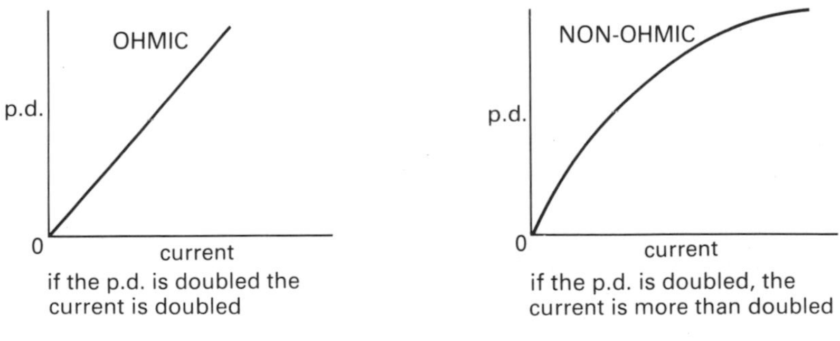

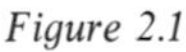
Figure 2.1

Circuit symbols

The symbols for 10 Ω and 4700 Ω resistors are shown in *figure 2.2*. These resistors have a single, fixed value (but it is sometimes necessary to introduce a resistor of variable value into a circuit). There are two types of variable resistor as shown in *figure 2.3*.

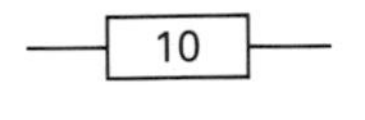

Figure 2.2
Symbol for a fixed resistor

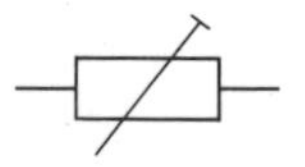

pre-set type adjusted by screwdriver

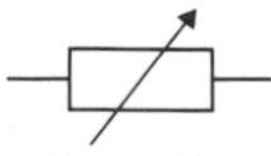

adjusted by spindle and knob

Figure 2.3 Symbols for variable resistors

Resistors in series and parallel

The combined effect of any number of resistors in **series** is found by **adding** their separate values.

Example Three resistors of values 10 Ω, 15 Ω and 12 Ω in series is equivalent to a single resistor of value 37 Ω.

The combined resistance R of **two** resistors R_1 and R_2 in **parallel** is found by working out

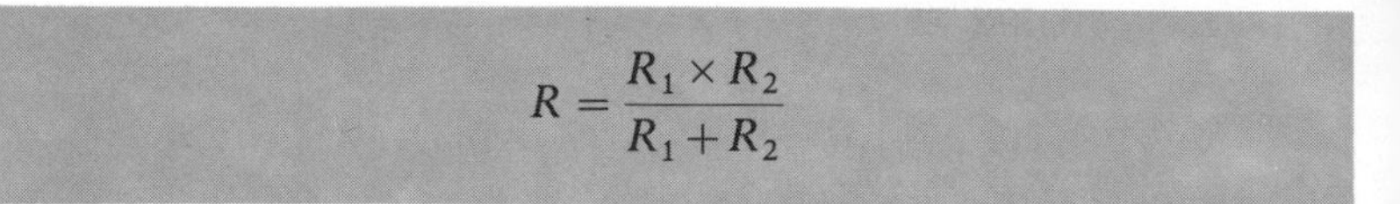

$$R = \frac{R_1 \times R_2}{R_1 + R_2}$$

Example Resistors of value 3 Ω and 6 Ω in parallel have a combined value of:

$$R = \frac{3 \times 6}{3+6}$$

$$R = \frac{18}{9}$$

$$R = 2\,\Omega$$

Only if you have more than two resistors in parallel need you use the equation:

$$\frac{1}{R} = \frac{1}{R_1} + \frac{1}{R_2} + \frac{1}{R_3} + \cdots$$

Notice that the combined resistance of any number of resistors in parallel is always less than the smallest value present. This is because there are more possible paths along which the current can flow.

Power rating of resistors

The maximum current which can be passed through a resistor, without it being damaged by overheating, can be calculated if its power rating is known. This figure gives the rate at which heat energy can be safely given out. Energy is measured in joules, so that the rate at which it is being given out would be measured in joules per second, or watts (W)

power in watts = rate of energy release in joules per second

If a current of I amperes passes through a resistance of R ohms, the energy W released every second is I^2R watts.

$$W = I^2R$$

As $V = IR$, then:

$$W = VI$$

Example 1 Calculate the maximum safe current through a $100\,\Omega$ 4 W resistor

$$W = I^2R$$

$$4 = I^2 \times 100$$

$$I^2 = \frac{4}{100}$$

$$\therefore\ I = \frac{2}{10} = 0{\cdot}2\,\text{A}$$

Example 2 If a resistor 'drops' 20 V at a current of 100 mA, calculatc its resistance and the minimum power rating. (Note: if a resistor has a p.d. across it of 20 V, it can be said to be producing a voltage drop of 20.)

first using:

$$V = IR$$

$$20 = 0{\cdot}1R$$

$$(100\,\text{mA} = 0{\cdot}1\,\text{A})$$

$$\therefore\quad R = 200\,\Omega$$

then using:

$$W = I^2R$$

$$W = (0{\cdot}1)^2 \times 200$$

$$= 0{\cdot}01 \times 200$$

$$\therefore\quad W = 2\,\text{W}$$

Commercial resistors and their applications

There are four types of resistor construction commonly met:

Carbon composition

These are the cheapest resistors and are intended for general purpose use in the power range 0·125 W to 1 W. They do, however, suffer from two main faults, namely their poor stability when the temperature changes, and their tendency to introduce 'noise' into the circuits in which they are used. The word 'noise' is used here to indicate the unwanted voltages generated by agitated electrons. If the equipment is working at audible frequencies, this noise can be heard as a rushing sound in the loudspeaker. For this reason, they have been almost completely replaced in new equipment by the carbon film type, which is now no more expensive, except where a resistor without inductance is required.

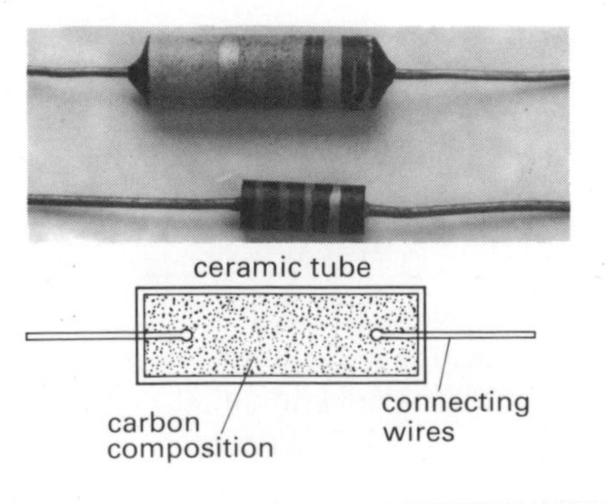

Left: Carbon composition resistors

Figure 2.4 Structure of carbon composition resistor

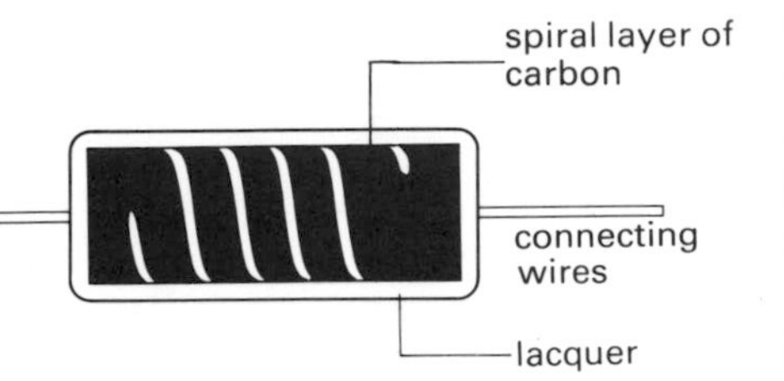

Figure 2.5 Structure of carbon film resistor

Carbon film

These consist of a hard, crystalline carbon film deposited on the outsic of a ceramic rod. They have a much better temperature stability an lower noise than carbon composition and are available in the pow range 0·25 W to 2 W.

Metal oxide

These have a similar construction to the carbon film type, with th carbon replaced by tin oxide. These resistors are used where maximur reliability and stability are required. They cost more than carbon filr resistors and are only readily available up to 0·5 W power rating.

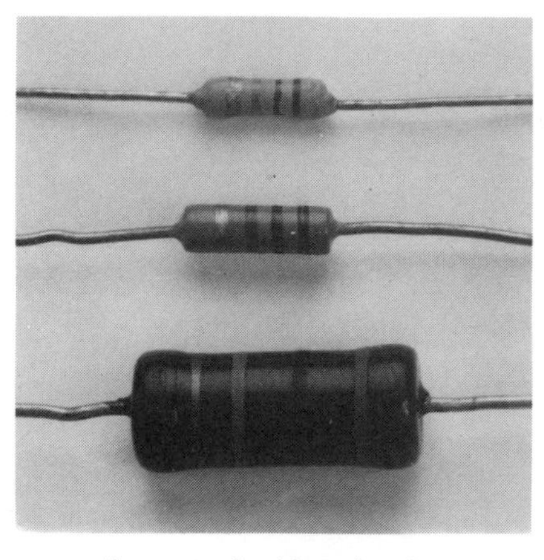

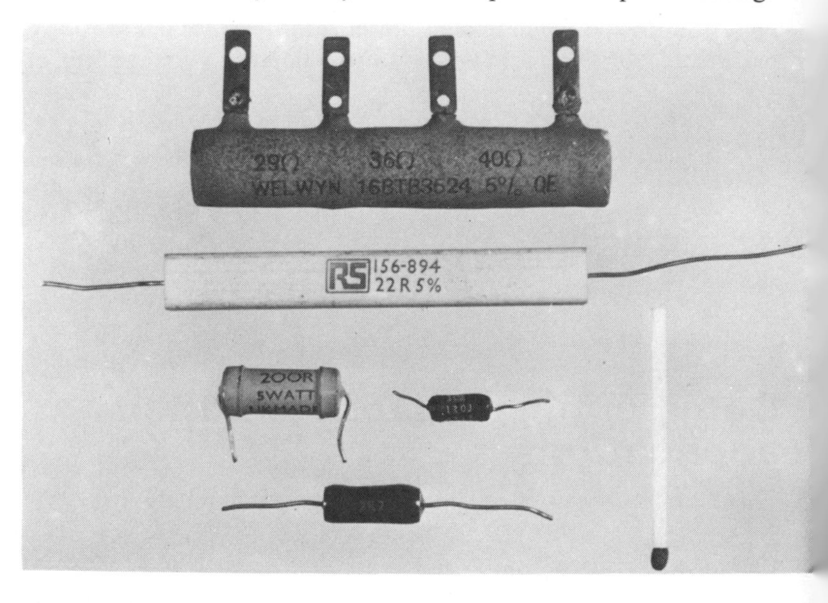

Above: metal oxide and carbon film resistors

Right: wirewound resistors

Wirewound

The resistance material here is a specially designed alloy wire, such as manganin, the resistance of which does not change much with temperature. The wire is wound on the outside of a ceramic tube and the whole assembly covered with high temperature enamel. Wirewound resistors fall into two groups:

a precision resistors, the values of which are known to within 0·1%.
b high power resistors capable of dissipating 25 W or more.

The British Standard 1852 code

This system is now being used to mark the value of resistors on circuit diagrams. The following examples show how it is used:

R47 means 0·47 ohm	100R means 100 ohm
1R0 means 1 ohm	1K2 means 1·2 kilohm
4R7 means 4·7 ohm	6K8 means 6·8 kilohm
47R means 47 ohm	2M7 means 2·7 megohm

No resistor can be expected to have a resistance exactly equal to the value written on it. The **tolerance** of a resistor is the percentage by which the resistor may be higher or lower than the value marked on it.

Example A 100 Ω resistor of 10% tolerance may have a value as low as 90 Ω or as high as 110 Ω.

On the B.S. 1852 code the following letters are used for tolerance F = ±1% G = ±2% J = ±5% K = ±10% M = ±20%.

Example 390RJ is 390 Ω ± 5%.

Preferred values

In order to avoid the need for an impossibly large number of resistor values having to be manufactured, only certain values are made. These values are called **preferred values.** The number of values in the series will depend upon the tolerance required, in order to avoid possible overlapping of adjacent values. Two series are used, E12 for 10% tolerance and E24 for 5% tolerance resistors.

The E12 series

10	12	15	18	22	27	33	39	47	56	68	82
100	120	150	180	220	270	330	390	470	560	680	820
1K	1K2	1K5	1K8	2K2	2K7	3K3	3K9	4K7	5K6	6K8	8K2

etc

The E24 series has these values **in addition** to those in the E12 series:

11	13	16	20	24	30	36	43	51	62	75	91
110	130	160	200	240	300	360	430	510	620	750	910
1K1	1K3	1K6	2K0	2K4	3K0	3K6	4K3	5K1	6K2	7K5	9K1

etc

The resistance colour code

In order to make the value of resistors easier to read, they carry a colour code consisting of 4 coloured bands (*figure 2.6*). The following colours are used:

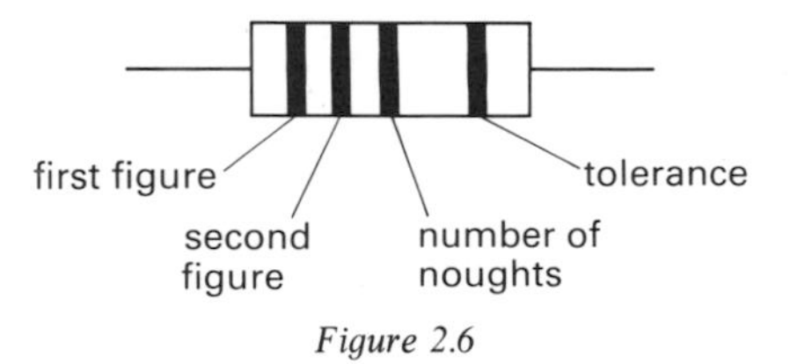

Figure 2.6

0 black
1 brown
2 red
3 orange
4 yellow
5 green
6 blue
7 violet
8 grey
9 white

tolerance band:
none ± 20%
silver ± 10%
gold ± 5%
red ± 2%

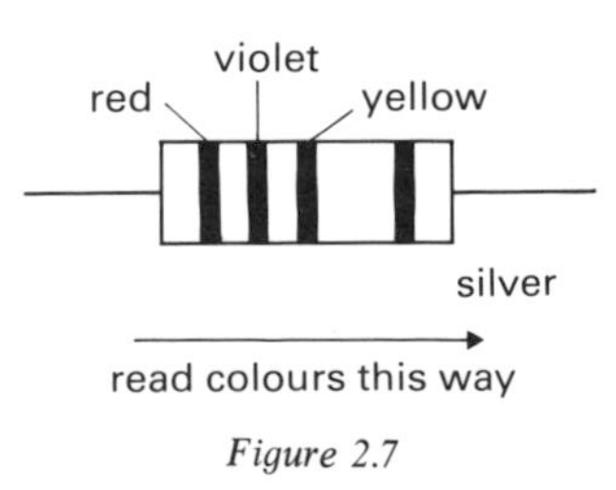

Figure 2.7

In the example (*figure 2.7*) the first figure is 2 (red), the second figure is 7 (violet) and the number of noughts is four (yellow). The tolerance is ±10% and the value given by the code is 270 000 Ω or 270 kΩ. If its value was measured accurately, it should not be more than 27 kΩ higher or lower than this value. Although this code may seem very slow to read at first, it should soon be possible to spot the pattern of colours for the common resistance values without having to stop and work them out. It is also useful to notice that the colours from 2 to 7 are in the same order as the rainbow.

Variable resistors

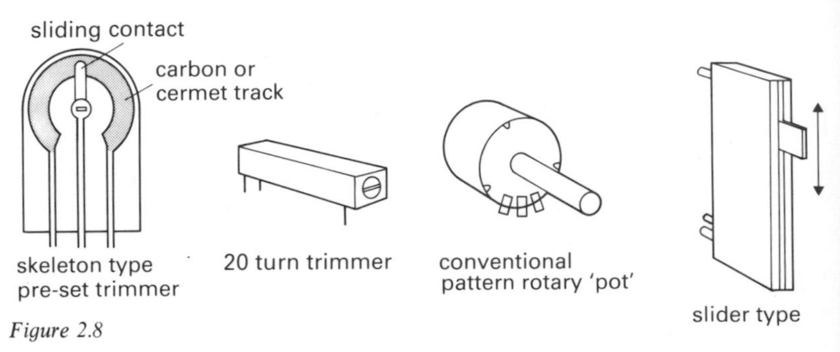

Figure 2.8

Figure 2.8 shows four types of variable resistor in common use. The track material for the trimmers may be carbon, or cermet (**cer**amic and **met**al oxide) for greater life and reliability. The track material for the conventional 'pot' may be carbon for ratings up to 1 W and wirewound for 1 W and above. Both slider and conventional 'pots' may be constructed in pairs for simultaneous operation in stereo circuits. The word 'pot' should be considered as an abbreviation for potential divider and not for potentiometer, as is often suggested. (For an explanation of the action of a potential divider see page 30.)

Some special types of resistor

Three types of non-ohmic resistor are often met in electronic circuits.

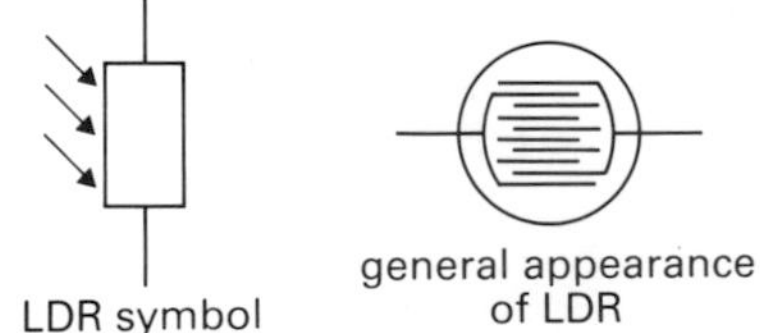

Figure 2.9

Light dependent resistors (or LDR)

This is also called a cadmium sulphide (Cd S) cell, but in fact is not really a cell because it is not a source of e.m.f. The resistance of an LDR falls considerably when exposed to light. The same effect is produced with radiation just outside the visible spectrum (infra red and ultra violet).

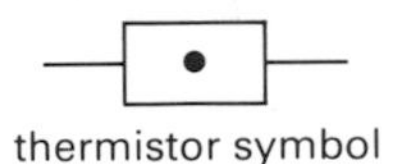

Figure 2.10

Thermistors

These have a resistance which varies considerably with temperature. The most common type are n.t.c. (negative temperature coefficient) which means their resistance falls as the temperature rises. The material from which they are constructed consists of oxides of manganese and nickel, with small quantities of copper, iron and cobalt to vary the properties as required. Large thermistors are used to prevent current 'surges' when certain electrical appliances are switched on. Small thermistors can be used for sensitive temperature measurement and control.

Voltage dependent resistors (VDR)

As their name suggests, these have a resistance value which can be altered by the p.d. across them. They are made by pressing silicon carbide with a ceramic binder and are used as surge limiters and voltage stabilisers.

Figure 2.11 VDR symbol

Internal resistance and the maximum power theorem

All power supplies have some resistance themselves; this is known as their **internal resistance**. In order to calculate the current which a power supply can drive through a given load, we need to know the e.m.f. and the internal resistance of the power supply as well as the resistance of the load connected to it (*figure 2.12*). The following equation can then be applied.

$$E = I(R+b)$$

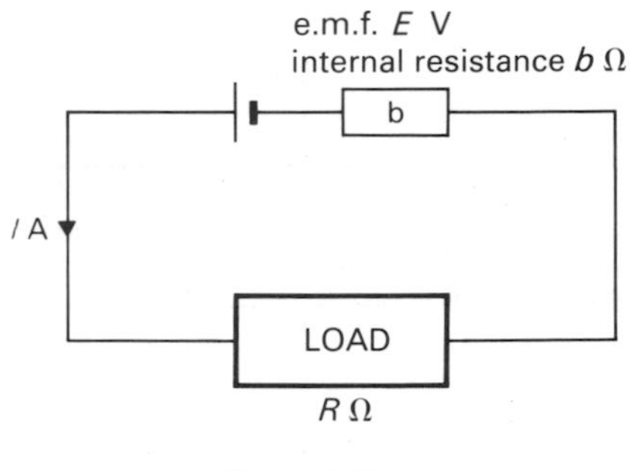

Figure 2.12

The maximum power theorem says, quite simply, that the greatest possible power is delivered to a load when the resistance of the load is equal to the internal resistance of the power supply (that is, when $R = b$).

Example A cell has an e.m.f. of 1·5 V and an internal resistance of 0·5 Ω. Calculate:

a the current it will drive through a load of 2 Ω.

b The power developed in a 2 Ω load.

c The maximum power the cell can deliver to a suitable load.

a

$$E = I(R+b)$$
$$1{\cdot}5 = I(2+0{\cdot}5)$$
$$I = \frac{1{\cdot}5}{2{\cdot}5} = 0.6\text{ A}$$

b

$$W = I^2R$$
$$= (0{\cdot}6)^2 \times 2$$
$$= 0{\cdot}36 \times 2$$
$$\therefore \text{ power} = 0.72\text{ W}$$

c maximum power is delivered into a load of 0·5 Ω

$$\therefore \quad 1{\cdot}5 = I(0{\cdot}5+0{\cdot}5)$$
$$I = \frac{1{\cdot}5}{1{\cdot}0}$$
$$I = 1{\cdot}5\text{ A}$$
$$W = (1{\cdot}5)^2 \times 0{\cdot}5$$
$$= 2{\cdot}25 \times 0{\cdot}5$$
$$\therefore \text{ power} = 1{\cdot}125\text{ W}$$

Unit 3 Capacitance and inductance

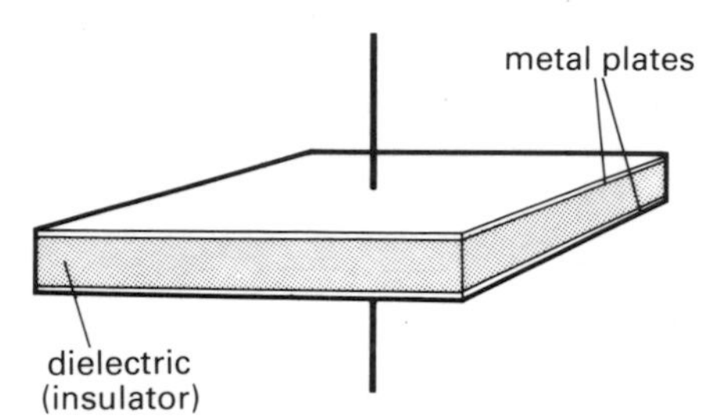

a. simplified construction of a capacitor

The first practical capacitor was constructed, almost accidentally, when in 1746 a Dutch physicist tried to store electric charge in a bottle of water. The action of the Leyden jar, as the bottle came to be called, was explained by Benjamin Franklin, who reconstructed the capacitor in its now familiar parallel plate form as shown in *figure 3.1a.*

If a battery, or some other source of potential difference is connected across the plates of a capacitor, there is a short drift of electrons from one plate to the other, see *figure 3.1b*. The capacitor is now said to be **charged**, one plate positively and the other plate negatively. There will be a p.d. across its terminals equal to the battery voltage. When the battery is disconnected, as in *figure 3.1c*, the capacitor will remain charged until the charge leaks away. This could take a long time, hence the danger of touching charged capacitors in recently operated high voltage equipment. If the plates of a charged capacitor are now joined by a conductor as in *figure 3.1d*, electrons will flow in the reverse direction until the p.d. across the plates has fallen to zero. Notice that at no time do electrons flow across the dielectric between the plates.

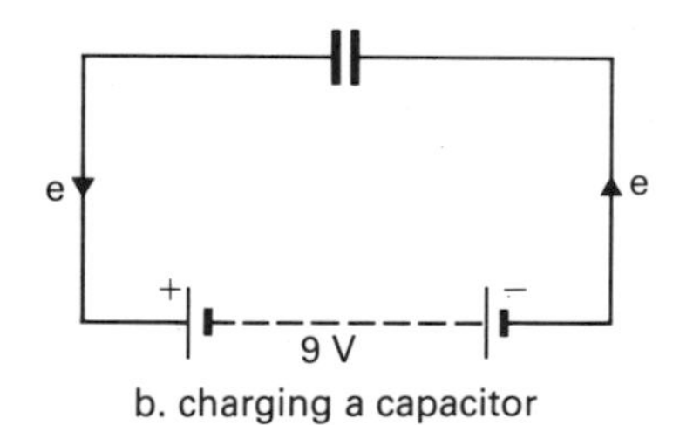

b. charging a capacitor

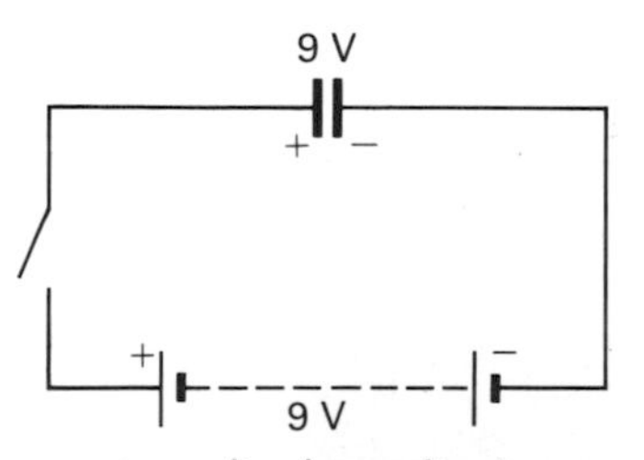

c. a capacitor keeps its charge

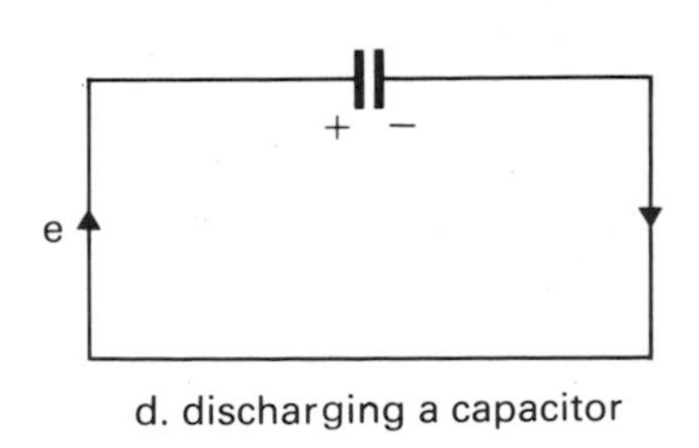

d. discharging a capacitor

Figure 3.1

The unit of capacitance

If a charge of 1 coulomb is given to a capacitor of 1 farad, the p.d. across its terminals will rise by 1 volt. If a charge of Q coulombs is given to a capacitor of C farads, and the resulting rise in p.d. is V volts, then the following equation applies:

$$Q = CV$$

The farad is a very large unit of capacitance, so for convenience, the following units are used:

microfarad μF ($= 10^{-6}$F)
nanofarad nF ($= 10^{-9}$F)
picofarad pF ($= 10^{-12}$F)

It is also probably worth remembering that

$1000\,\text{pF} = 1\,\text{nF}$

Circuit symbols

Two circuit symbols are necessary as there are two quite different types of capacitor construction (*figure 3.2*). The **non-polarised** capacitor can be connected either way round in a circuit, but the **polarised** one must be connected according to the signs marked on it. If the signs are not visible, the shape of most wire-ended electrolytic capacitors will give the information (*figure 3.3*). Notice also that a capacitor is marked not only with the value of its capacitance, but also with a voltage, to indicate the highest p.d. which should be connected across it. If this p.d. is exceeded, the dielectric layer may be damaged.

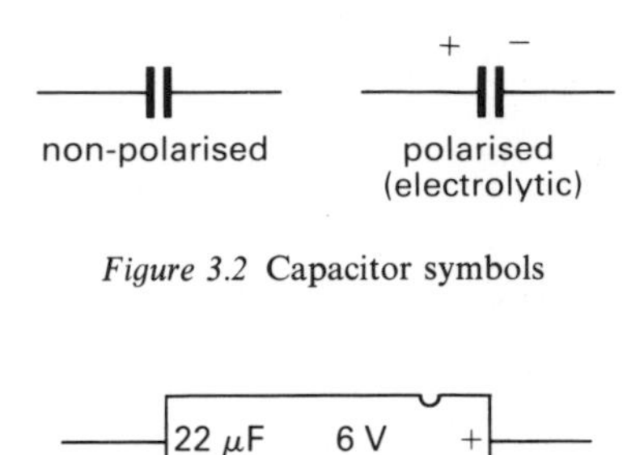

Figure 3.2 Capacitor symbols

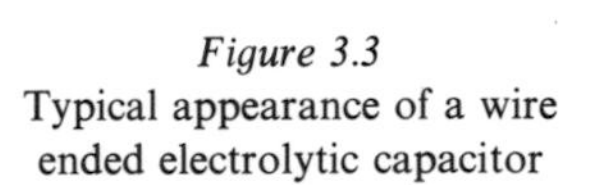

Figure 3.3
Typical appearance of a wire ended electrolytic capacitor

Different types of capacitor

Electrolytic

Figure 3.4 shows the 'swiss roll' construction of most electrolytic capacitors. These are available in a range of values from 1 μF to 50 000 μF. The large values are made possible by using a very thin dielectric layer, consisting of an oxide layer on the aluminium or tantalum foil electrodes. In aluminium electrolytics, the two aluminium electrodes are separated by a layer of porous paper, soaked in ethylene glycol (ethandiol).

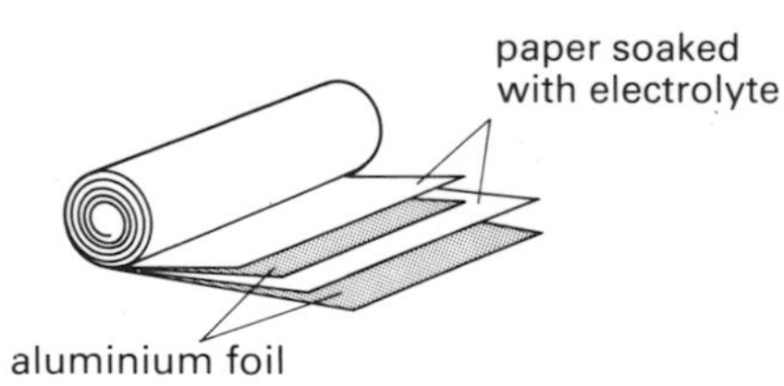

Figure 3.4 Construction of capacitors

Electrolytic capacitors are only suitable for low frequency work, up to about 10 kHz.

Tantalum electrolytics are much smaller than the corresponding aluminium type, but are much more expensive.

It is unwise to use very old electrolytic capacitors as they become 'leaky' with age (see testing capacitors, p. 21).

Above: Precision waxed paper capacitors

Paper

The insulating material between the aluminium foil electrodes consists of waxed paper. The range of values is usually 0·001 μF to 8 μF and they are suitable for medium frequency work.

Plastic film

Four types of plastic may be used as the insulating layer between the aluminium foil electrodes. They are:

polypropylene polyester polycarbonate polystyrene

With the exception of polystyrene, which has a low melting point, the other plastics may have a metallic film deposited on them by a vacuum evaporation process. This is what is meant by **metallised** capacitors. The range of values available depends upon the type of construction and metallised polyester can be made up to 10 μF, but polyester film/foil are made up to 0·01 μF.

Silvered mica

These rather expensive capacitors are made by depositing a thin layer of silver on each side of a thin sheet of mica.

The normal range of values is 1 pF to 0·01 μF and they are very suitable for high frequency work. They also have excellent stability and are accurate to within 1% of the marked value.

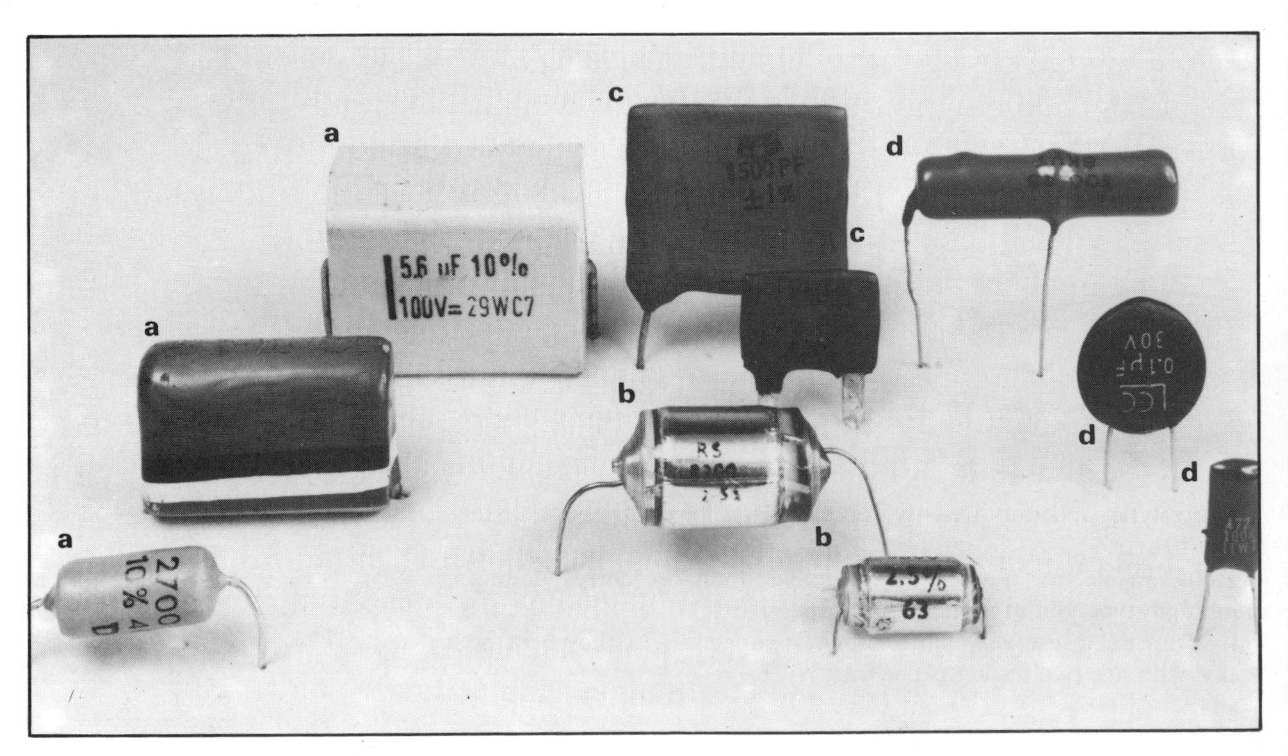

Below: Various types of capacitor

a polyester
b polystyrene
c silver mica
d ceramic

Ceramic

These capacitors consist of a silver plated ceramic tube or disc. They can be made to give a relatively large capacitance in a small space and are suitable for high frequency work.

Air spaced

These capacitors are usually constructed to have a variable value by altering the amount of overlap between two or more metal plates. The dielectric in this case is the air between the plates. These capacitors are often used as the tuning control inside a radio receiver.

Variable air spaced capacitor

Capacitors in parallel

To find the combined effect of any number of capacitors in **parallel**, **add** their values together. The voltage at which the combination can work will be decided by the lowest working voltage capacitor present.

Example The three capacitors in *figure 3.5* have values of $0{\cdot}1\ \mu F$, $0{\cdot}22\ \mu F$ and $0{\cdot}047\ \mu F$.

$$\text{total capacitance} = 0{\cdot}1 + 0{\cdot}22 + 0{\cdot}047 = 0{\cdot}367\ \mu F$$

The working voltage will be 160 V.

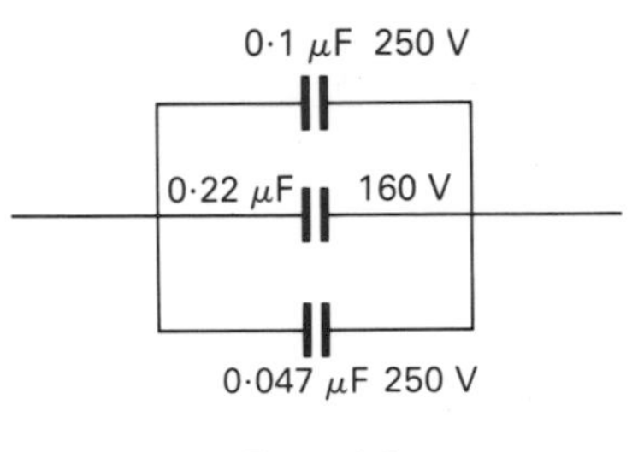

Figure 3.5

Capacitors in series

For two capacitors, C_1 and C_2, in **series** use this formula:

$$C = \frac{C_1 \times C_2}{C_1 + C_2}$$

For more than two capacitors in series, use the equation:

$$\frac{1}{C} = \frac{1}{C_1} + \frac{1}{C_2} + \frac{1}{C_3} + \cdots$$

The simplest way of finding the working voltage of several capacitors in series is to arrange for them all, individually, to have equal working voltages. The working voltage of the series combination is then the sum of the individual values.

Example If two capacitors of value $0{\cdot}1\ \mu F$ and $0{\cdot}22\ \mu F$, both of working voltage 160 V, are wired in series (*figure 3.6*) the capacitance of the combination will be

$$\frac{0{\cdot}1 \times 0{\cdot}22}{0{\cdot}1 + 0{\cdot}22} = \frac{0{\cdot}022}{0{\cdot}32} = 0{\cdot}069\ \mu F$$

and the working voltage will be 320 V.

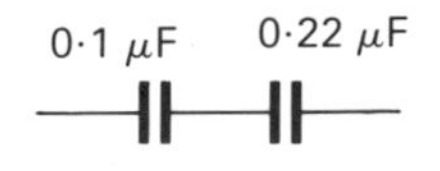

Figure 3.6

Applications of capacitors

1 Capacitors can block d.c. but allow a.c. to pass (*figure 3.7*).

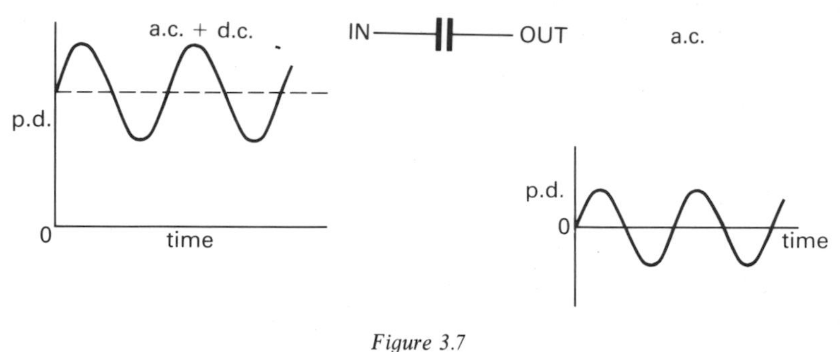

Figure 3.7

2 Capacitors can smooth rectified a.c. to make d.c. (*figure 3.8*).

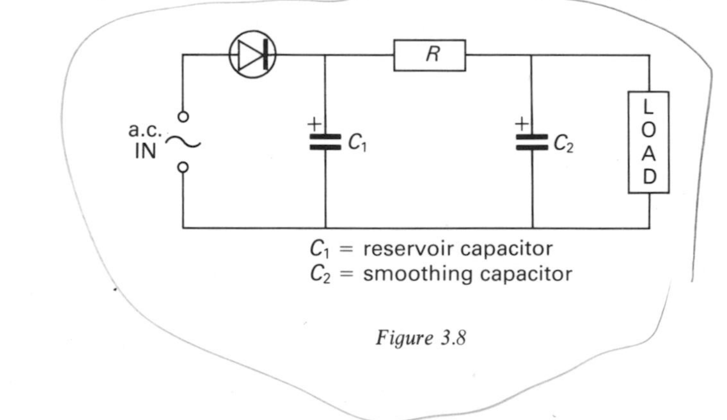

Figure 3.8

3 Capacitors can act as a 'wattless' resistor in a.c. circuits. The resistance to the flow of a.c. is called the **reactance** of a capacitor and is given the symbol X_c. We can calculate its value by using the equation:

$$X_c = \frac{1}{2\pi f C}\ \Omega$$

where f = frequency of the a.c. in hertz
C = capacitance in farads

Example The reactance of a 0·1 μF capacitor at 50 Hz

$$= \frac{1}{2\pi \times 50 \times 0{\cdot}1 \times 10^{-6}}$$

$$= \frac{10^6}{10\pi}$$

$$= 31{\cdot}8\ \text{k}\Omega$$

Notice that as C and f increase, so X_c decreases, as shown by *figure 3.9*.

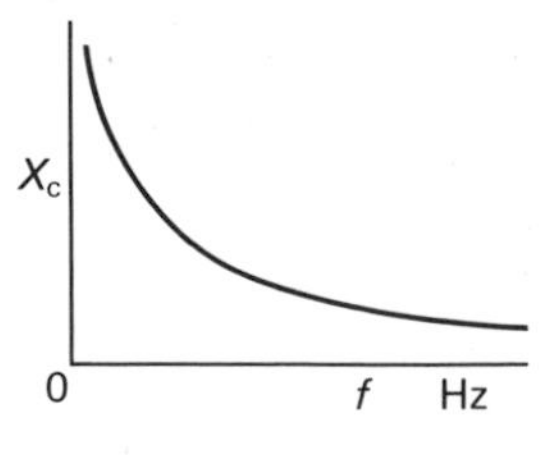

Figure 3.9 Graph to show how the reactance of a capacitor varies with frequency

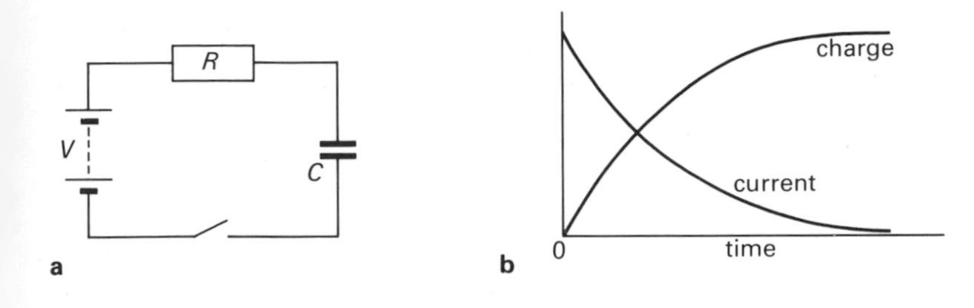

Figure 3.10a Charging a capacitor through a resistor

Figure 3.10b Graphs to show how charge and current vary with time

4 Capacitors can be used with a resistor to act as a time constant (*figure 3.10a*). A heavy current flows when the switch is first closed. Then, as the charge builds up, so the current falls (*figure 3.10*b). The maximum final charge Q coulombs which builds up on the capacitor is equal to CV coulombs, where C is the capacitance in farads and V is the potential difference across it in volts.

$$Q = CV$$

The time taken for the charge to rise to 63% of its final value is equal to CR seconds. The value of CR is called the **time constant** of the circuit.

Example If $C = 4\,\mu\text{F}$ and $R = 3\,\text{M}\Omega$

$$CR = 4 \times 10^{-6} \times 3 \times 10^{6}$$
$$= 12\,\text{s}$$

It should also be noted that the capacitor will be fully charged after about $5CR$ seconds. Hence, in the above example, the capacitor would take about 60 seconds to charge up to the p.d. of the battery.

The reverse process also works, so if a resistor of R ohms is connected across a capacitor of C farads carrying a charge of Q coulombs, the charge falls by 63% in CR seconds and reaches zero in about $5CR$ seconds (*figure 3.11*).

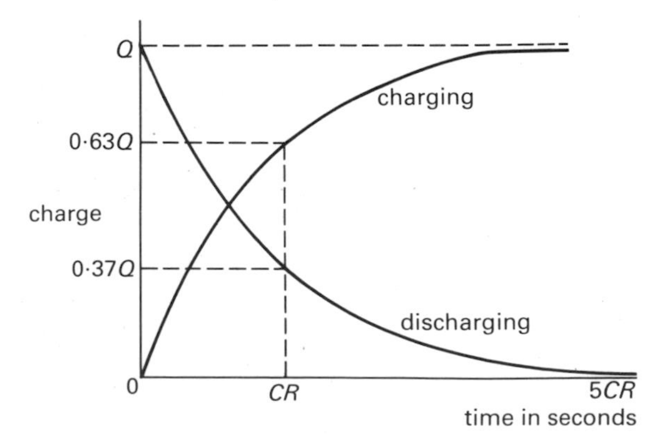

Figure 3.11
Graphs to show the time taken for a capacitor to charge and discharge

5 Capacitors may be used to store charge, as in a photographic electronic flash unit.

Stored charge in coulombs $= CV$

Stored energy $= \frac{1}{2}CV^2$ joules

$$W = \tfrac{1}{2}CV^2$$

$C =$ capacitance in farads

$V =$ p.d. in volts

Example Calculate the value of capacitor required to store 100 J at a potential of 200 V.

$$100 = \tfrac{1}{2} \times C \times (200)^2$$

$$200 = 40000C$$

$$C = \frac{2}{400}\,\text{F}$$

$$C = \frac{2 \times 10^6}{400}\,\mu\text{F}$$

$$= 0{\cdot}5 \times 10^4\,\mu\text{F}$$

$$\therefore \quad C = 5000\,\mu\text{F}$$

Capacitor in an electronic flash unit

Testing capacitors

A multimeter set to its high resistance measurement range can be used to test capacitors. A capacitor should not pass d.c., so if its resistance is much less than 1 megohm, it can be said to be 'leaking'. When testing large values of non-polarised capacitors, the meter needle may be seen to 'kick' briefly as the capacitor charges up. When testing electrolytic capacitors, two points must be remembered:

a The negative terminal of the meter (the black wire) is usually positive on its resistance range, so make certain it is connected to the positive side of the capacitor (this is usually identified by a + sign).

b An electrolytic capacitor must have a p.d. correctly applied to it if the insulating dielectric is to be formed. When the meter is first connected, the resistance will be quite low, but will soon rise as the dielectric forms.

Inductance

When a current flows through a wire, a magnetic field is created around that wire. When the wire is wound into a coil, the magnetic field passes through the turns of the coil. If the magnetic field changes in strength around this coil, an e.m.f. is set up in that coil. The direction of this e.m.f. is always such as to oppose the change which produced it. (This is called **Lenz's law**.)

Thus if a current in a coil is changing in strength, the resulting change in magnetic field sets up a **back e.m.f.** which tends to oppose the changing current. This property of a coil is called **(self) inductance**. The unit of inductance is the **henry (H)**.

The inductance of a coil is 1 henry if the current changing in it at the rate of 1 ampere per second produces an e.m.f. of 1 volt.

The henry is quite a large unit, and the millihenry (mH) is commonly met.

air cored

iron cored

dust iron cored

Figure 3.12
Circuit symbols for inductors

How the current grows in an inductor

In the circuit shown in *figure 3.13*, the current rises from zero to 63% of its final steady value in a time of L/R seconds. The value of L/R is called the **time constant** of the circuit. If the current has reached its steady value and the switch is opened, it falls by 63% in L/R seconds. At the same time, the rapidly changing magnetic field sets up a high back e.m.f. and this causes a spark to appear at the switch contacts.

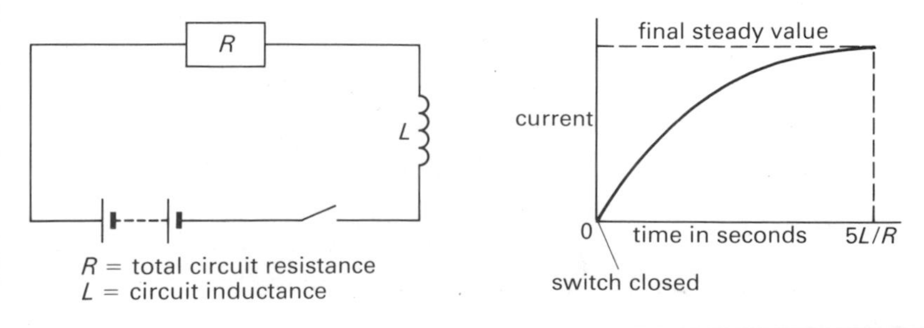

Figure 3.13 The growth of current through an inductor. Note that the value of the final steady current depends only upon the total resistance of the circuit and the battery voltage.

Reactance of an inductor

In a d.c. circuit, an inductor has a small resistance made up only from the slight resistance of the wire from which it is wound. In a circuit with a.c. flowing, the constantly changing current is opposed not only by this resistance, but also by the self inductance of the coil. This opposition offered by the self inductance is called the **reactance** of the inductor and its value can be calculated by using the equation:

$$X_L = 2\pi f L$$

where X_L is the reactance in ohms
f is the frequency in hertz
L is the inductance in henrys

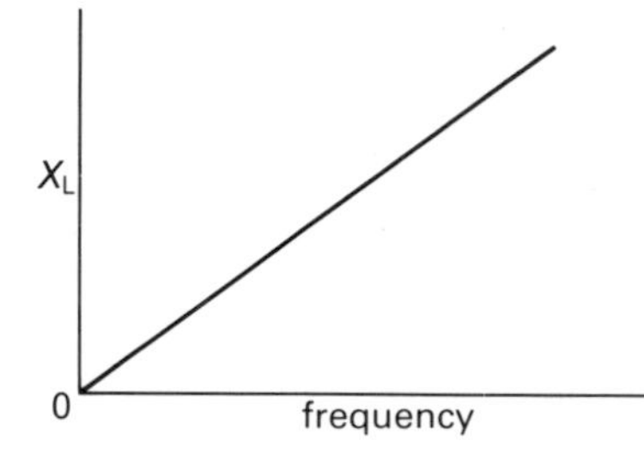

Figure 3.14
Graph to show how the reactance of an inductor changes with frequency

How the reactance of an inductor changes with frequency

As the frequency of the current through an inductor rises, so does the reactance, as shown in *figure 3.14*. Compare this graph with *figure 3.9* which shows the graph for a capacitor.

Since both inductors and capacitors have reactance, it is sometimes useful to distinguish between them by calling them **inductive reactance** and **capacitative reactance** respectively.

Inductor types and uses

Large value iron cored inductors

These are also called **chokes**. They may be used in two different ways:

a to control current in a.c. circuits, for example in a fluorescent tube (*figure 3.15*). The great advantage of using an inductor to control the current instead of a resistor is that no heat is generated in the inductor, apart from that produced by the resistance of its windings.

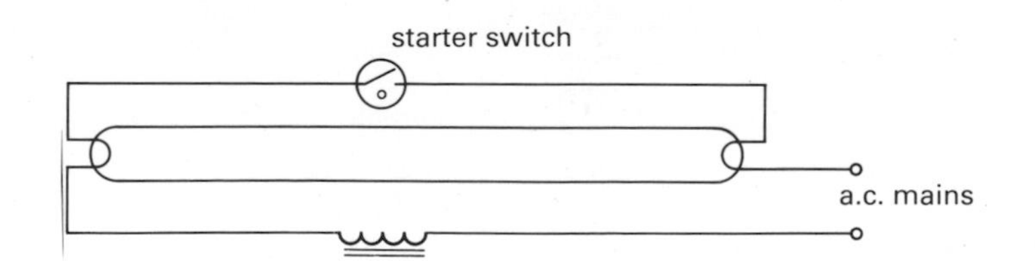

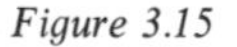

Figure 3.15

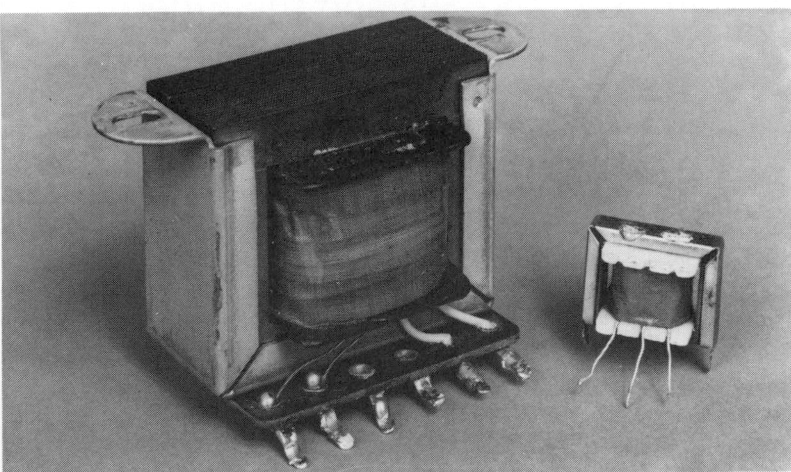

Large and small value inductors

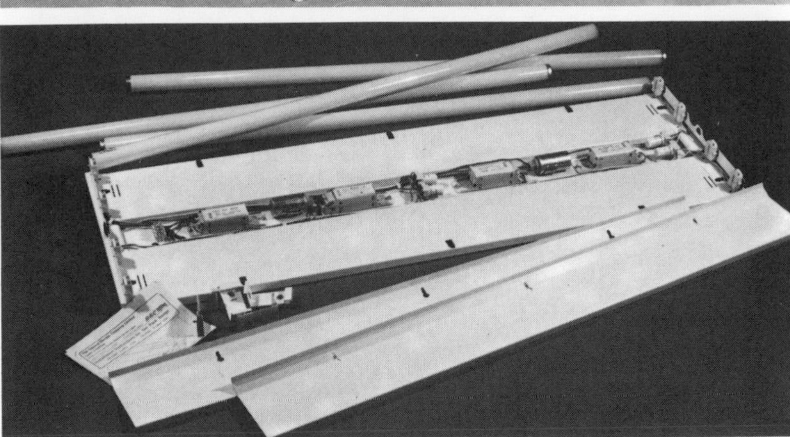

Chokes and capacitors in a fluorescent tube unit

b to help produce smooth d.c. from rectified a.c. (*figure 3.16*). The inductor is much more effective than a resistor in the same position (*figure 3.8*).

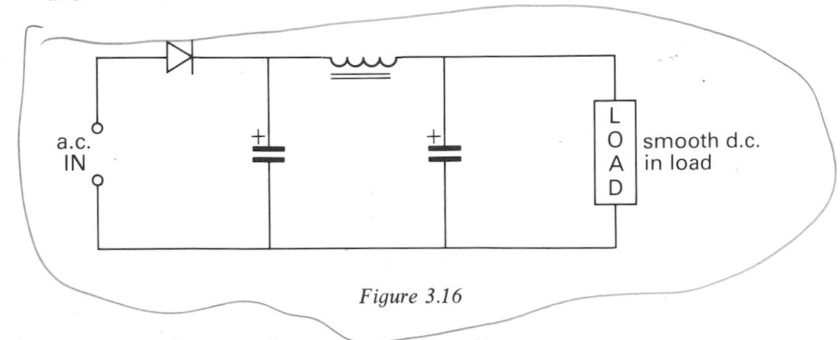

Figure 3.16

Small value air cored coils

These are used as radio frequency chokes; they choke back, or oppose the passage of very high frequency a.c. They can be used to remove high frequency signals from a circuit in which both high and low frequency signals are present. The small inductance offers little reactance to low frequencies but a high reactance to radio frequencies.

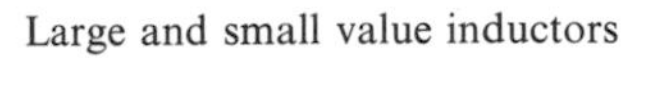

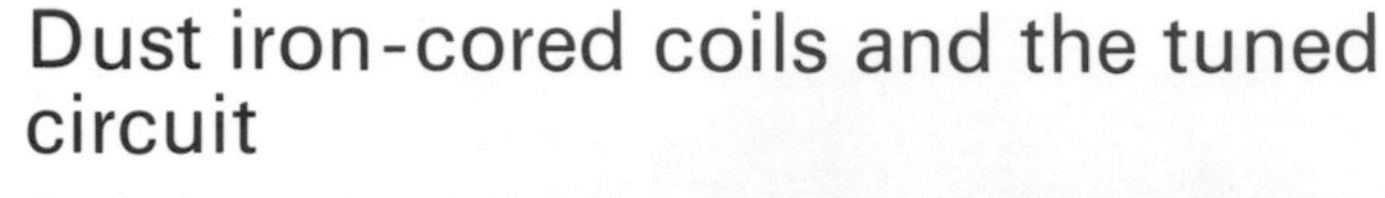

Dust iron-cored coils and the tuned circuit

Both air-cored and dust iron-cored coils may be used in a **tuned circuit**, such as may be found in a radio receiver (see unit eighteen). Dust iron cores are made by mixing iron powder with a non-conducting binder; they are very fragile because this mixture is very brittle. The result is a core which is magnetic but will not allow electric currents (called eddy currents) to flow in it. The eddy currents would produce unwanted energy losses. A coil can be made to have a variable inductance by arranging for the core to be screwed in and out of its centre.

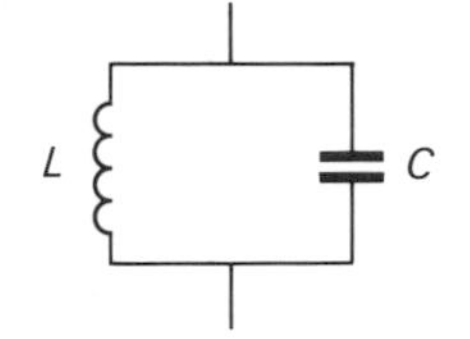

Figure 3.17
A parallel tuned circuit

A tuned circuit is shown in *figure 3.17*. At a frequency called the **resonant frequency** (f_R) it has its lowest reactance. The frequency at which this occurs can be varied if either C or L are variable and the value of f_R can be calculated by using the equation

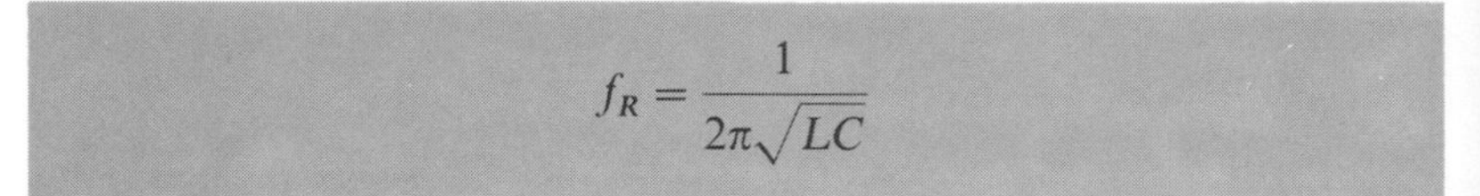

$$f_R = \frac{1}{2\pi\sqrt{LC}}$$

The graph in *figure 3.18* shows how the reactance of the capacitor and inductor change with frequency.

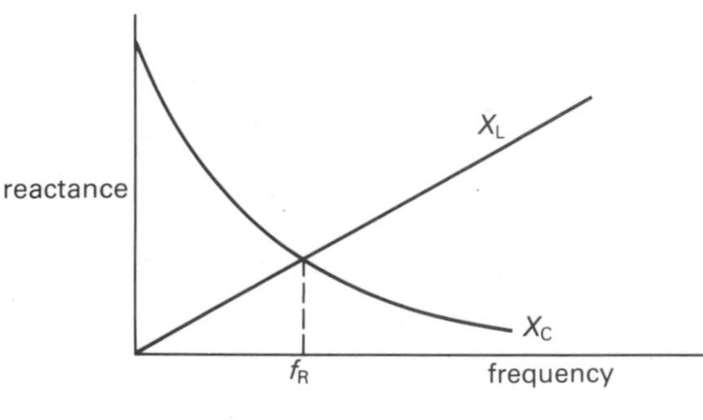

Figure 3.18

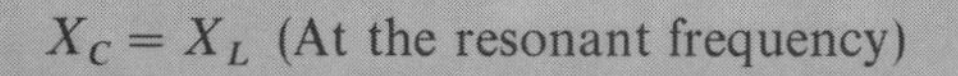

$$X_C = X_L \text{ (At the resonant frequency)}$$

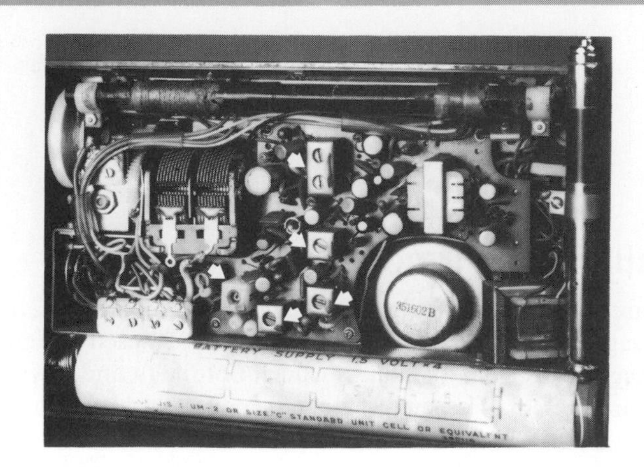

Dust iron-cored coils (arrowed) in a radio receiver

Measuring current and voltage

Unit 4

One of the most useful and essential tools for the electronic engineer is a good quality testmeter. At the heart of this instrument is a **moving coil galvanometer**. This consists of a rectangular coil of wire, pivoted so that it can turn in a field produced by a strong magnet. Fixed to this coil is a long pointer which moves over a scale. A screwdriver slot on the front enables a small adjustment to be made, if necessary, to make the pointer read zero when no current is flowing through the coil.

When a current does flow in the coil, it creates another magnetic field and the coil is made to turn. As it turns it winds up a spiral hair spring until the force produced by the current balances the force produced by the spring. The stronger the current, the further the coil turns, and so the scale of such a meter has divisions which are always uniformly spaced.

Moving coil meter movement

The moving coil and its magnet form what is known as the **basic movement**. The basic movement has two important electrical properties:

a the resistance of the coil.

b the current required to move the pointer from zero to the far end of the scale. This is called the full scale deflection (F.S.D.) current.

Example If the coil has a resistance of 100 Ω and is moved to F.S.D. by a current of 50 μA, Ohm's law can be applied to calculate the p.d. across the coil:

$$V = IR$$

$$V = 50 \times 10^{-6} \times 100$$
$$= 5 \times 10^{-3}$$

$$V = 5\,\text{mV}$$

This means that a p.d. of 5 mV connected across the coil will also produce F.S.D. Thus the basic movement described above is both a microammeter, reading up to 50 μA, and a millivoltmeter reading up to 5 mV.

Extending the range of a voltmeter

The basic movement described above can be made to give readings when used on p.ds. above 5 mV by using a resistor in *series* with it. A resistor used in this way is called a **multiplier**.

Example If the meter is to give F.S.D. with 10 V, then 10 V must drive a current of 50×10^{-6}A through the coil. Use Ohm's law to calculate the **total** resistance:

$$R = \frac{V}{I}$$

$$R = \frac{10}{50 \times 10^{-6}}$$

$$R = 200\,000\,\Omega$$

The coil has a resistance of 100 Ω, so a resistor of value 199 900 Ω must be used as the multiplier. Multipliers always tend to have 'awkward' or non-preferred values. Sometimes they are available ready made, otherwise, the above multiplier could be made from the preferred values 180 kΩ + 18 kΩ + 1·8 kΩ + 100 Ω.

The quality of a voltmeter

The quality of a voltmeter is usually taken to mean its sensitivity as a measuring instrument. The more sensitive it is, the less current is required to move the pointer to F.S.D. A perfect voltmeter would require no current to operate it; a good voltmeter is one which requires a very small operating current. The sensitivity is usually expressed in **ohms per volt** and is calculated as follows:

$$\text{sensitivity} = \frac{\text{total resistance of meter + multiplier}}{\text{p.d. required to give F.S.D.}}$$

Thus, for the meter reading 10 V at F.S.D.

$$\text{sensitivity} = \frac{200\,000}{10}$$
$$= 20\ \text{k}\Omega/\text{V}$$

For work with transistors a sensitivity of 100 kΩ/V is recommended, if the voltmeter current is not to interfere with the circuit being tested.

Extending the range of an ammeter

The current range is extended by placing a resistor in **parallel** with the moving coil. This resistor is called a **shunt**. If the previous basic movement is to read F.S.D. with 1 A then 0·000 05 A will pass through the meter (50 μA) and 0·999 95 A must bypass the meter through the shunt (*figure* 4.1).

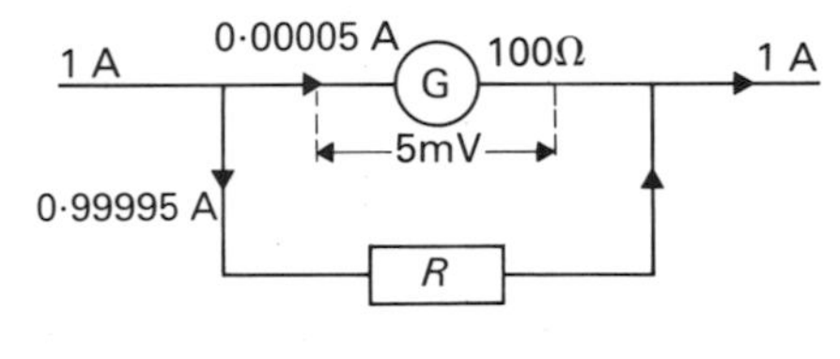

Figure 4.1

Left: An ammeter with its shunts

As can be seen from the diagram, the p.d. across both the shunt and the coil is 0.005 V (5 mV). Using Ohm's law on the shunt:

$$V = IR$$
$$0{\cdot}005 = 0{\cdot}999\,95 \times R$$
$$R = \frac{0{\cdot}005}{0{\cdot}999\,95}$$
$$R = 0{\cdot}005\,000\,2\ \Omega$$

Notice that the shunt has a very **low** resistance. When an ammeter is placed in series with a circuit, it should not add much resistance to that circuit. The perfect ammeter would not increase the resistance of the circuit in which it is connected.

Types of meter

The universal testmeter

A universal testmeter, such as the Avometer below provides measurement over a wide range of currents and voltages for both d.c. and a.c. up to a frequency of 2 kHz. In addition to this, there is a scale for the direct measurement of resistance.

Such a meter may have a mirror scale to eliminate parallax errors. To use the mirror you should line up the reflection of the pointer immediately under the pointer itself. When the pointer is covering its reflection, your eye is in the correct position for the scale to be read.

If the size of the voltage or current to be measured is not known, even roughly, the meter should be set to the highest available voltage or current range. The range can then be set lower to give the highest possible scale reading.

Notice the + mark on the meter terminal. It is this terminal, usually coloured red, which should be connected to the positive side of the voltage or current being measured. *Figure 4.2* shows the way a voltmeter and an ammeter should be connected in a circuit.

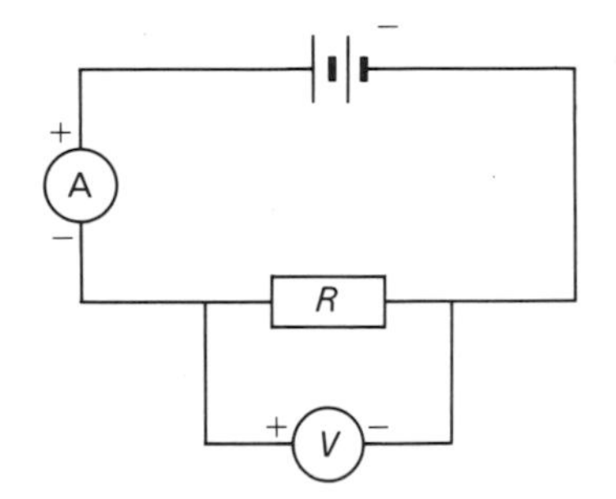

Figure 4.2 How to connect an ammeter and a voltmeter into a circuit

An Avometer

Overload protection is provided on most high quality meters. It can take many different forms, and in some Avometers an excess current through the meter causes a cut-out button to pop up. When this happens, the meter is automatically disconnected from the circuit which overloaded it.

The electrostatic voltmeter

This meter works as a result of the forces between electric charges (mentioned in unit one). No current flows through the meter when it is in use, but it is not easy to construct such a meter to give readings below about 100 V. They are most suitable for very high voltage measurement, such as the final anode potential of a television tube. The scale of an electrostatic meter is not uniform: the divisions are wider in the middle of the scale.

Moving iron ammeters

These tend to be cheap, robust instruments and are used where high accuracy is not important, for example to measure the charging current in a car battery charger. They work on both a.c. and d.c. and their scales are not uniform.

Hot wire ammeter

This type of meter works equally well on a.c. and d.c. since it relies upon the heat produced when current flows through a resistance. The main advantage of this meter is that it will continue to read a.c. correctly at very high frequencies.

Tongs ammeter

When very heavy currents are flowing through thick wire, it is not usually possible to insert an ammeter in the circuit. Because there is quite a strong magnetic field around such a wire, two soft iron jaws clamped across this conductor can carry the magnetic flux to a special meter, where it produces a deflection.

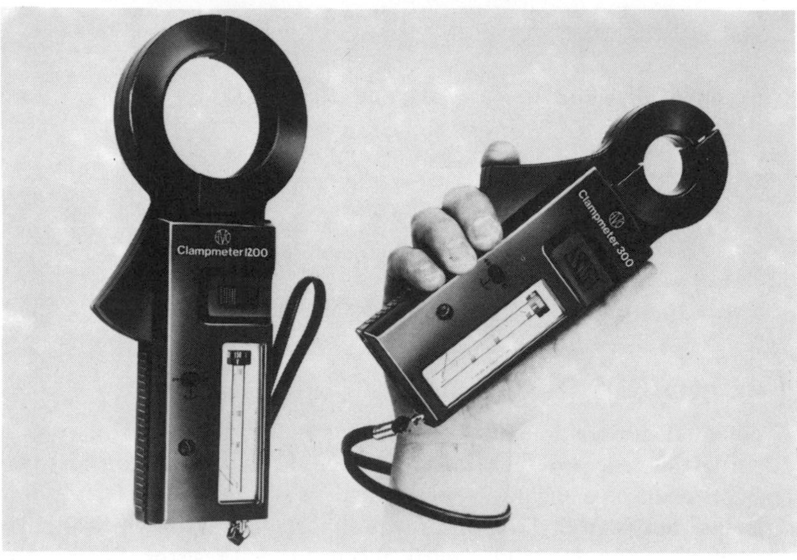

Tongs ammeter

A digital Avometer

Digital meter

Many students have difficulty in reading an ordinary meter with a pointer, especially if the scale has more than one set of numbers written on it. On digital meters, the reading required is displayed as a number and usually with very great accuracy.

The cathode ray oscilloscope (CRO)

The CRO has a built in amplifier which enables it to measure very small voltages. Its chief value is, however, to measure and display the 'shape' of electrical signals which vary with time. It can also measure the frequency of a.c. and its peak value. For further details on the operation of a CRO see unit six.

Circuit analysis

e.m.f. E volts; internal resistance b ohms

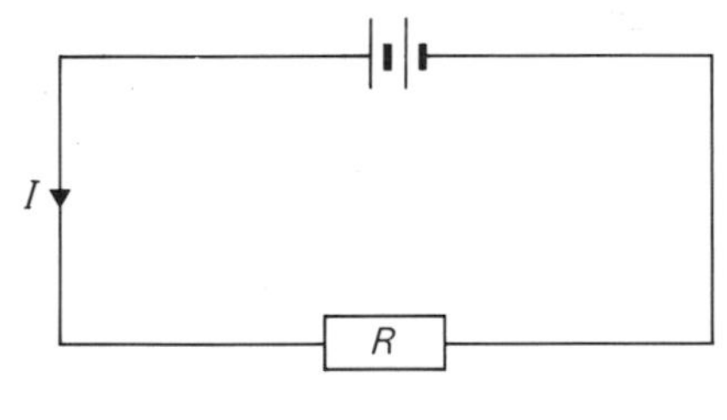

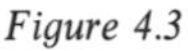
Figure 4.3

To calculate the current in a simple series circuit, you need to know the e.m.f. and total circuit resistance (*figure 4.3*).

$$E = I(R+b)$$

Example If $E = 6\,\text{V}$, $b = 2\,\Omega$ and $R = 10\,\Omega$

$$6 = I(10+2)$$
$$I = 0{\cdot}5\,\text{A}$$

To calculate the currents in a simple branching circuit such as *figure 4.4* use the following formula:

$$I_1 = I \times \frac{R_2}{R_1+R_2}$$

Example If $I = 2\,\text{A}$, $R_1 = 4\,\Omega$ and $R_2 = 6\,\Omega$

$$I_1 = 2 \times \frac{6}{10}$$
$$\therefore \quad I_1 = 1{\cdot}2\,\text{A}$$
$$\therefore \quad I_2 = 0{\cdot}8\,\text{A}$$

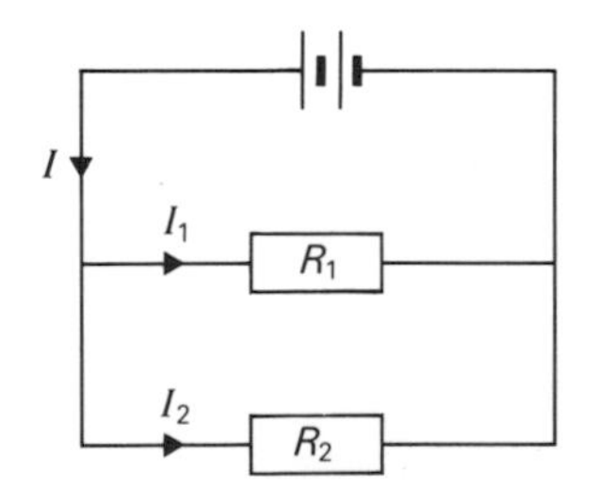

Figure 4.4

Notice that the larger current flows through the smaller resistor.

The potential divider

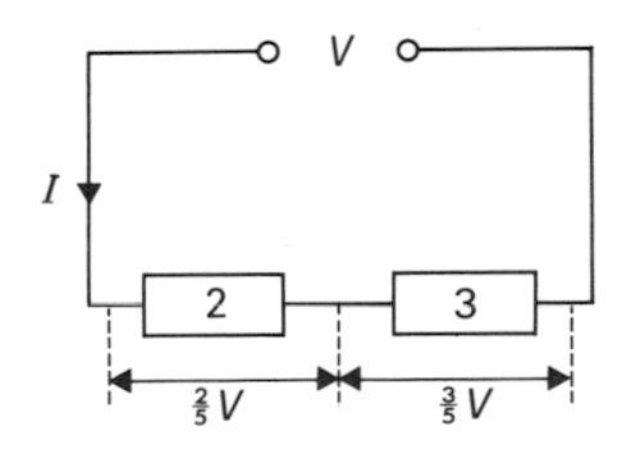

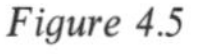
Figure 4.5

A potential divider does exactly what its name suggests; it divides a potential difference (voltage) into a number of equal parts. Consider two resistors connected across a source of p.d. as shown in *figure 4.5*. In this circuit, V has been divided into 5 equal parts $(2+3)$, of which 2 parts appear across the $2\,\Omega$, and 3 parts across the $3\,\Omega$ resistor. Exactly the same result would be obtained for resistors of value $2\,\text{k}\Omega$ and $3\,\text{k}\Omega$, because it is their ratio which determines the result. There is, however, one important difference between the low value and the high value pairs, namely their effect upon the circuit current I. It is essential to calculate I and to make sure it will not produce too much heat in the chosen resistors. Unit one, page 9 shows how this can be done.

It is important to realise that if a potential divider is used to drive current through another circuit, the calculated value of p.d. will no longer be correct. The error will be slight, however, if the current being drawn from the divider is small compared with I. Or to put it another way, the second circuit must have a high resistance compared with the potential divider resistance across which it is connected. There is a further discussion of the potential divider in unit ten.

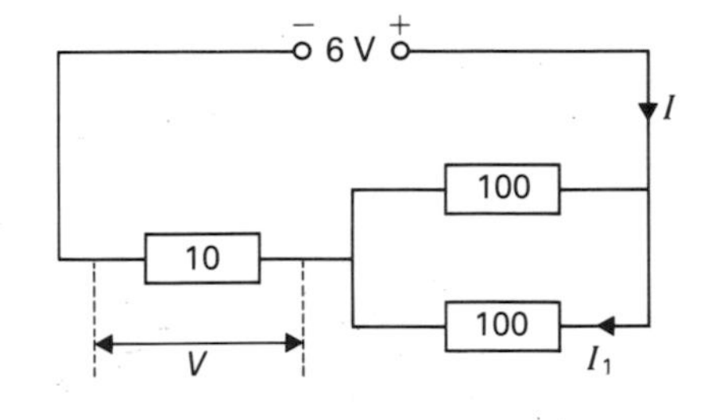

Figure 4.6

Example In *figure 4.6*, calculate I, V, and I_1. Because the two 100 Ω resistors in parallel act like a single 50 Ω resistor, the circuit can be redrawn as in *figure 4.7*.

Calculate the current using:

$$I = \frac{V}{R}$$

$$I = \frac{6}{10+50} = \frac{6}{60}$$

$$I = 0{\cdot}1 \text{ A}$$

Then calculate the p.d. across the 10 Ω resistor using $V = IR$

$$V = 0{\cdot}1 \times 10$$

$$V = 1 \text{ V}$$

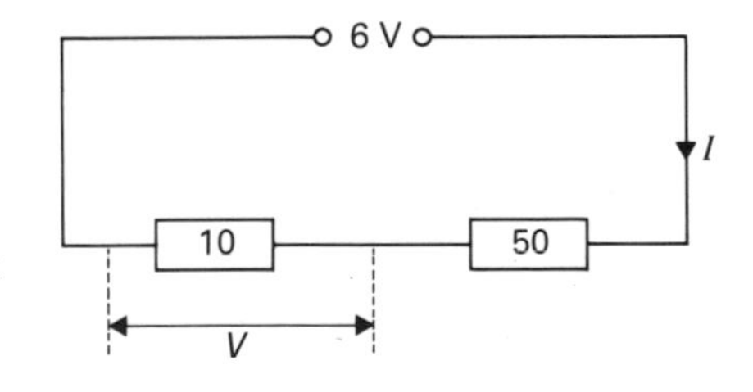

Figure 4.7

Now returning to *figure 4.7*, we find I_1, by dividing the current according to the equation:

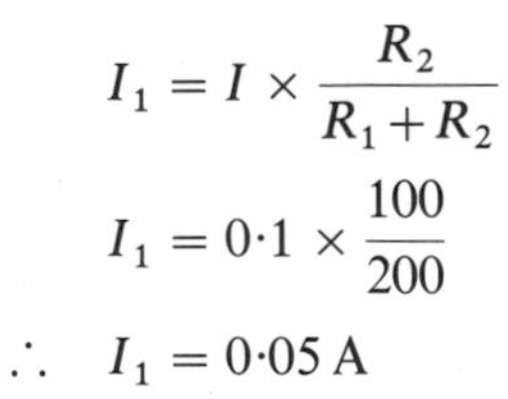

$$I_1 = I \times \frac{R_2}{R_1+R_2}$$

$$I_1 = 0{\cdot}1 \times \frac{100}{200}$$

$$\therefore \quad I_1 = 0{\cdot}05 \text{ A}$$

The Ohmmeter

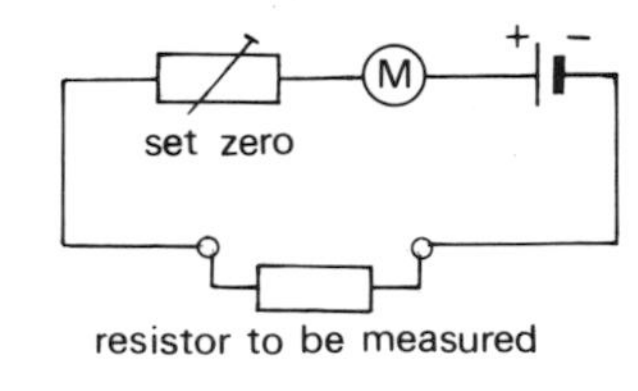

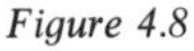

Figure 4.8

A quick and practical method of measuring resistance is to use an ohmmeter, or a multimeter set to its resistance range. The simplified circuit for an ohmmeter is shown in *figure 4.8*.

To use the meter, join the terminals with a wire and adjust the 'set zero' until the meter reads zero ohms. (This corresponds to maximum deflection.) Now remove the short circuit across the terminals and insert the resistor to be measured. The meter will give a reading of the resistance, provided that it has a correctly marked resistance scale, which is not uniform (*figure 4.9*).

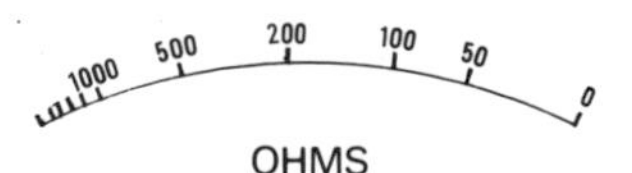

Figure 4.9

Unit 5

Alternating current and the transformer

Figure 5.1 shows the principle of the a.c. generator or alternator. In reality the magnetic field is provided not by permanent magnets, but by electromagnets. The armature would consist of more than one turn of wire on an iron core. For further details of the alternator see unit nine.

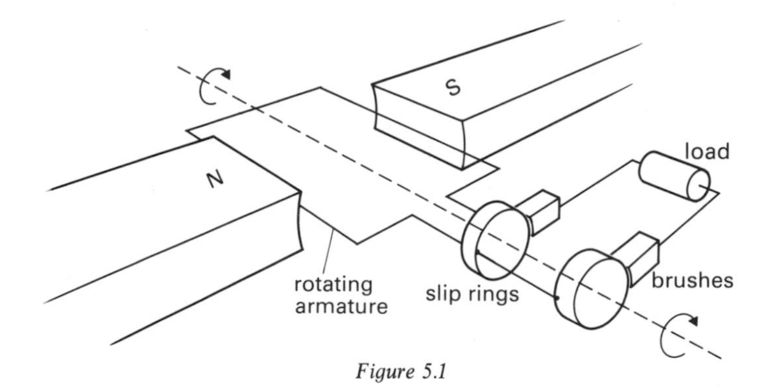

Figure 5.1

The a.c. cycle

Figure 5.2 shows how the output e.m.f. of an alternator varies with time. Two complete cycles are shown: each one corresponds to one complete revolution of the alternator and takes 0·02 seconds. The **frequency** of a.c. is the number of cycles completed in 1 second and the unit is hertz (Hz). The frequency shown on the graph would be 50 Hz, the normal frequency of a.c. mains in the U.K. The mains frequency in USA is 60 Hz.

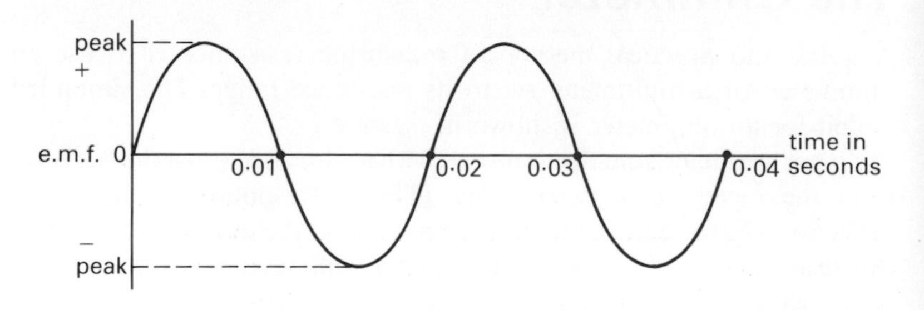

Figure 5.2

Peak r.m.s. and average values

As the e.m.f. and current are constantly changing in the a.c. cycles, use is made of what are known as **root mean square** (r.m.s.) values. An alternating current of, say, 5 A r.m.s. would produce exactly the same heating effect as 5 A d.c. The a.c. does, of course, rise above this value for part of the time and fall below it for the rest, but the heating effect produced is the same as that with 5 A d.c. If you need to know the peak value reached, use this equation:

$$\text{peak value} = \sqrt{2} \times \text{r.m.s. value}$$

Example The insulation on a 250 V r.m.s. mains cable has to withstand the peak e.m.f.

$$\begin{aligned}\text{peak e.m.f.} &= 1{\cdot}414 \times 250 \\ &= 353\ \text{V}\end{aligned}$$

A capacitor is sometimes found connected across a mains switch to reduce sparking when the switch is operated. This capacitor would be rated at 400 V d.c. in order for it to survive the peak e.m.f.

If an alternating current or e.m.f. has its value stated, it is always assumed to be the r.m.s. value unless otherwise stated.

The **average** value of one complete a.c. cycle is zero if the wave is symmetrical, that is if the shape of the positive half is the same as the negative half. Thus the average is usually taken to mean the average of one half of the a.c. cycle. The average value is very rarely needed and, if necessary, it is calculated using the equation:

$$\text{r.m.s. value} = 1{\cdot}11 \times \text{average value}$$

The relative positions of peak, r.m.s. and average values are shown in *figure 5.3*.

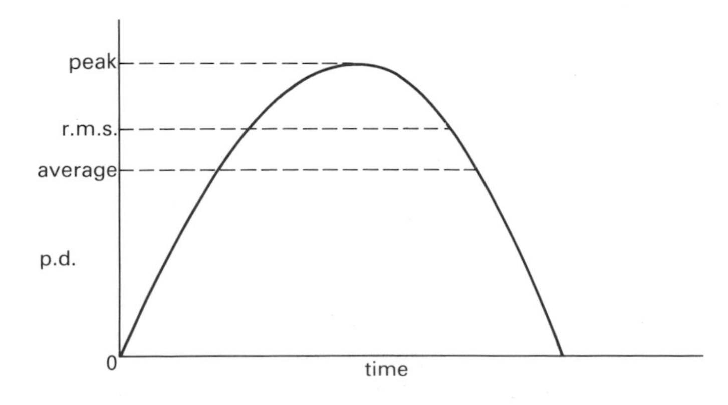

Figure 5.3

Transformers

The transmission of electricity over long distances is accompanied by a loss of power in the transmission wires of the National Grid. As there are about 7000 km of overhead and 1000 km of underground cable, these losses must be kept to a minimum. This is done by transmitting the power at a very high voltage. To see how this works, let us calculate the energy lost by sending 100 kW of power along a wire of resistance 0·01 Ω,

a at a pressure of 250 V and **b** at a pressure of 250 kV:

a using

$$W = VI$$
$$100\,000 = 250I$$
$$I = \frac{100\,000}{250}$$
$$I = 400 \text{ A}$$
$$\text{heat produced} = I^2R = (400)^2 \times 0{\cdot}01 = 1600 \text{ W}$$

Thus 1·6 kW of power is wasted as heat.

b

$$100\,000 = 250\,000I$$
$$I = 0{\cdot}4 \text{ A}$$
$$\text{heat produced} = (0{\cdot}4)^2 \times 0{\cdot}01 = 0{\cdot}0016 \text{ W}$$

At this higher pressure, only 1·6 mW of power is wasted.

In the United Kingdom the majority of the National Grid operates at 400 kV.

Since electricity cannot be generated at this pressure and would certainly not be safe to use, the action of transformers is used to step up and then step down again. The circuit symbol for these transformers is shown in *figure 5.4*.

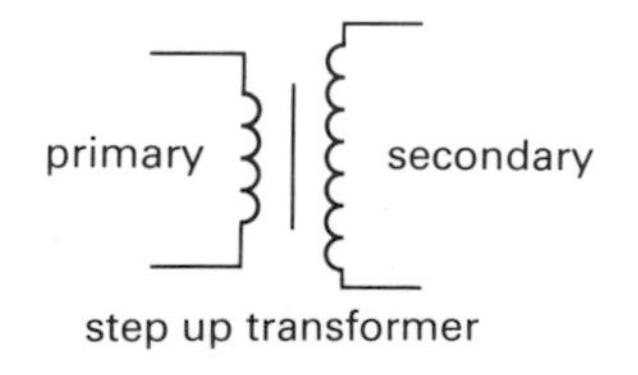

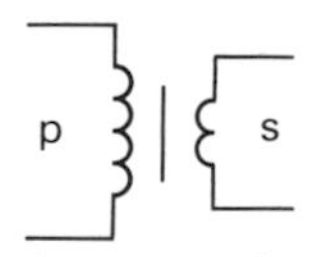

Figure 5.4
Circuit symbols for transformers

Right: A National Grid transformer at a substation in Worcestershire

Construction of transformers

A transformer consists of a **primary** coil, a **secondary** coil and a **core**. These can be arranged in different ways and one type is shown in *figure 5.5*. The core is not solid, but is made up from thin sheets or **laminations** of silicon steel, each one insulated from the other. This construction prevents eddy currents from flowing in the core and so wasting energy. If the transformer is to be used at radio frequencies, the core material will be dust iron or ferrite (another non-conducting magnetic material).

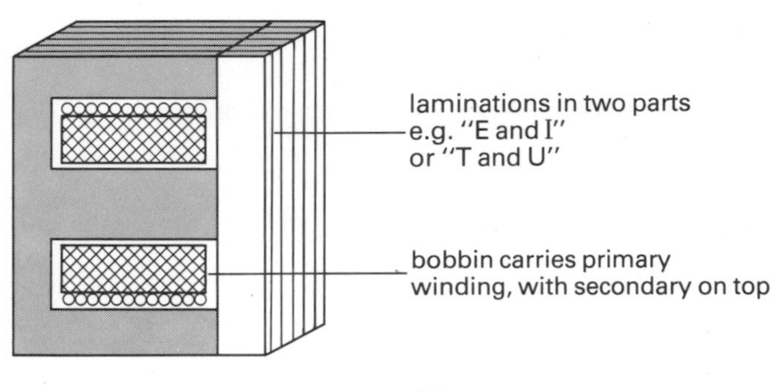

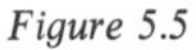
Figure 5.5

The turns ratio

If all of the magnetic flux produced by the primary passes through the secondary, then:

$$\frac{\text{number of turns on primary}}{\text{number of turns on secondary}} = \frac{\text{e.m.f. applied to primary}}{\text{e.m.f. generated at secondary}}$$

Example If the primary can deliver a maximum power of 100 W (or rather, 100 VA, because the primary does not release power but passes it to the secondary), this power is available at the secondary as, say, 2 V at 50 A, or 500 V at 200 mA. In the first case the turns ratio would be 250/2, and in the second it would be 250/500. The **actual number** of turns on the primary, however, would be nearer 2500 than 250.

Transformer efficiency

If a transformer is 100% efficient, all of the power released by the primary passes to the secondary, and no heat is produced in either the core or the windings. This is not possible in practice because the copper wire windings must have some resistance. A 90% efficiency for a small transformer is quite usual and for larger ones 98% is possible. The efficiency of a transformer can be measured by using voltmeters and ammeters, on their a.c. range, in the circuit shown in *figure 5.6*.

$$\text{efficiency} = \frac{\text{power out}}{\text{power in}} \times 100\%$$

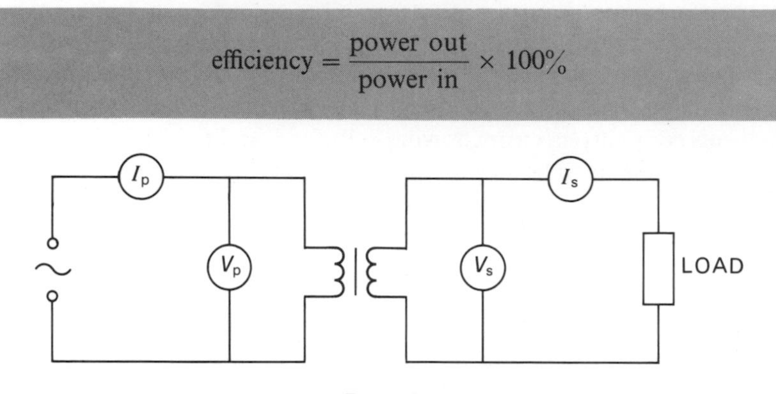

Figure 5.6
Circuit used to measure transformer efficiency

Example If $I_p = 120\,\text{mA}$ and $V_p = 250\,\text{V}$

$$\text{power in} = 250 \times 0{\cdot}12$$
$$= 30\,\text{W}$$

if $I_s = 2\,\text{A}$ and $V_s = 12\,\text{V}$

$$\text{power out} = 2 \times 12$$
$$= 24\,\text{W}$$

$$\text{efficiency} = \frac{24}{30} \times 100$$

$$\therefore \quad \text{efficiency} = 80\%$$

The transformer as a matching device

From the maximum power theorem (unit two), we know that the resistance (or impedance) of a load should equal that of its source. If an a.c. source of e.m.f. is to feed a load of different impedance from itself, a transformer may be used between them to match the two impedances. The turns ratio of this transformer is calculated as follows:

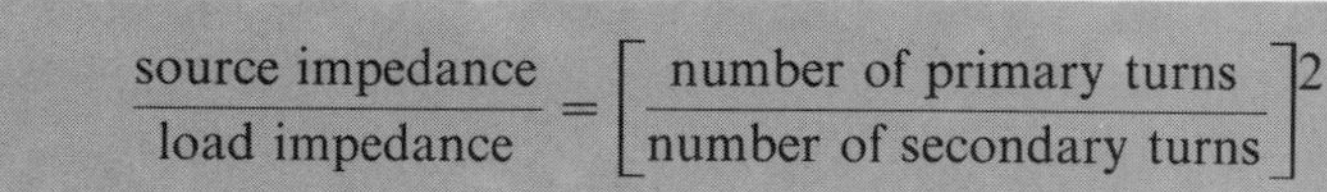

$$\frac{\text{source impedance}}{\text{load impedance}} = \left[\frac{\text{number of primary turns}}{\text{number of secondary turns}}\right]^2$$

Matching transformers may be found:

a in a valve amplifier to match the low impedance loudspeaker to the high impedance valve which drives it.

b in early transistor radios and amplifiers to match the collector impedance to the lower input impedance of the next stage (see unit eleven).

c matching a low impedance ribbon or dynamic microphone to the high impedance input of an amplifier (unit nine).

The autotransformer

If the secondary does not need to be electrically isolated from the primary, an autotransformer may be used. This is just a single winding with tappings (*figure* 5.7).

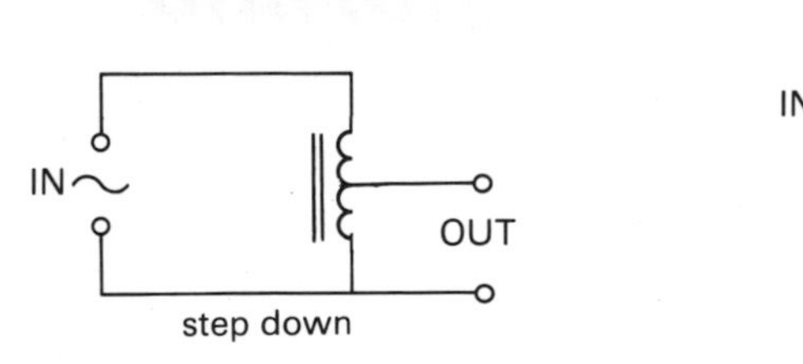

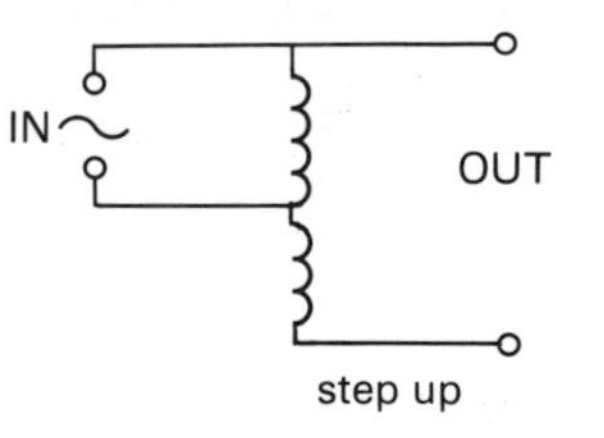

Figure 5.7

Unit 6

The cathode ray tube and oscilloscope

There are two electronic devices in common use which employ a cathode ray tube (CRT):

a the cathode ray oscilloscope, which uses a tube with electrostatic deflection.

b the television receiver, which uses a tube with magnetic deflection.

The electron beam is produced and focussed by means of a series of electrodes known as the **electron gun**, which is common to both tubes.

The electron gun

The simplified construction of the two types of tube is shown in *figure 6.1*. The tungsten filament, operating well below the temperature of a tungsten filament lamp, releases very few electrons. Its alternative name, the heater, indicates its true purpose, namely to raise the temperature of the electron emitting surface of the cathode. The cathode consists of a nickel cylinder coated at its tip with a substance designed to release electrons at dull red heat. The rate at which electrons are being released could be decreased by lowering the temperature of the cathode, but this would be too sluggish for most practical purposes. The intensity of the electron stream is controlled by surrounding the cathode with a metal cylinder charged negatively with respect to the cathode. This cylinder is called the **grid** (a name borrowed from the age of thermionic valves), and the more negative it is made, the more strongly it repels the electrons, so causing fewer to emerge through its central hole. Maximum electron flow is obtained when the grid is at the same potential as the cathode.

The electron beam is then made to accelerate towards the screen by causing it to pass through a series of cylindrical electrodes, which are charged positively with respect to the cathode. These electrodes are called the **anodes** and the highest positive voltage would be found on the final anode, which consists of a graphite coating on the inside of the glass. In the cathode ray oscilloscope, the p.d. between the final anode and the cathode is around 6 kV, whereas in a colour television tube it is around 25 kV.

Since the electron beam is required to produce a small luminous spot when it strikes the phosphor coating, the beam needs to be shaped or **focussed** in order to have a small cross-sectional area as it reaches the end of the tube. This focussing is carried out by altering the voltage on one of the first anodes.

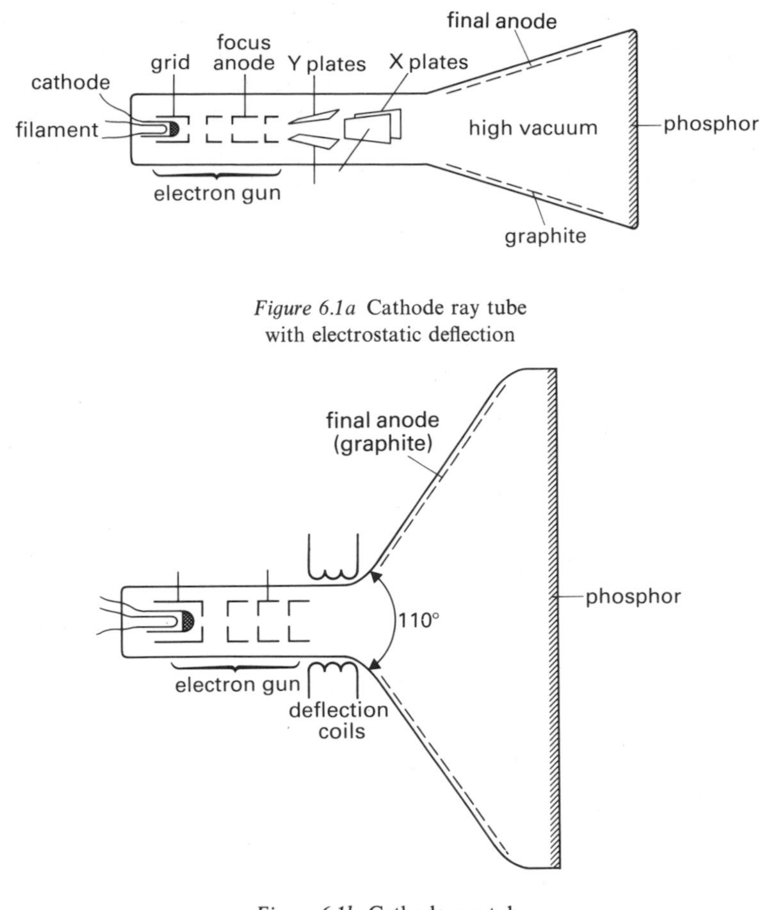

Figure 6.1a Cathode ray tube with electrostatic deflection

Figure 6.1b Cathode ray tube with electromagnetic deflection

Having struck the screen, the electrons are brought to rest, and the secondary electrons they produce are collected by the final anode, and return to the cathode via the power supply. So, the electrons flow in a complete circuit, just as they would if only solid conductors were present.

Deflecting the electrons

In both types of cathode ray tube the electron beam can be made to move vertically and horizontally. In a cathode ray oscilloscope the vertical deflection is produced by means of an electrostatic field between two metal plates, called the *Y* plates (because they cause deflection along the *Y* axis). The electron beam will move towards the plate which is made positive and away from the negative plate. The beam must not strike these plates, so they are not parallel but angled to allow maximum deflection. A similar electrostatic field between the *X* plates causes deflection along the *X* axis. The *Y* plates are always the ones furthest from the screen, because they have a greater influence upon the beam than the *X* plates. Can you think of some reasons for this? Try to imagine the effect of the electron velocity and the beam length on the movement of the spot by each pair of plates charged to a given p.d.

The very large angle of deflection required for a television tube can only be obtained by using a magnetic field. One pair of coils produces a magnetic field which gives the *X* deflection and another pair provide the *Y* deflection. These are called the line coils and the field (or frame) coils respectively. A 25 inch T.V. tube (that is a tube having a screen with diagonal 63cm) would need to be 1 metre long if it made use of electrostatic deflection.

The cathode ray oscilloscope

A cathode ray oscilloscope (CRO or 'scope) is found in most electronic workshops and laboratories. As a good one is very expensive, it must be an important instrument to be found so frequently. It is certainly the most versatile of all electronic instruments. The three main uses for a CRO are:

a to measure p.d. – both d.c. and a.c. up to 10 MHz or more.

b to measure the frequency of a.c. signals.

c to study the 'shape' of a.c. signals – that is to see how their voltage varies with time.

There are other, more accurate instruments for operations **a** and **b**, so it is for operation **c** that the oscilloscope is unique.

Figure 6.2 shows the principal controls to be found on a CRO, but their positions may vary from those shown. The control labelled **time base** is responsible for adjusting the speed at which the spot is drawn across the screen. When the time base is turned off, a stationary spot can be centred on the graticule.

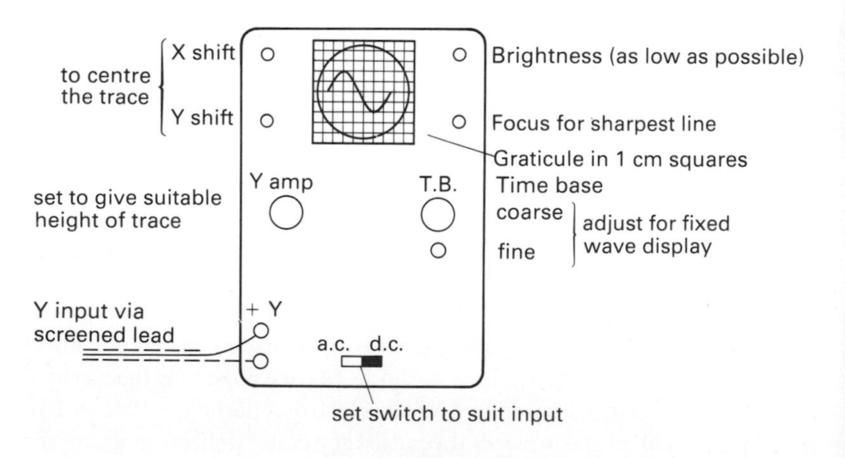

Figure 6.2
The main controls on an oscilloscope

If a d.c. signal is connected to the *Y* input terminals of a CRO, with the time base turned off, the spot will move to a new place on the screen either above or below its previous position. (The spot will move up if the positive input terminal is connected to the positive side of the d.c. source and down if connected to the negative side.) On an oscilloscope with a calibrated Y input, the p.d. required to move the spot 1cm will be marked. *Figure 6.3* shows the effect of connecting a 1.5 V cell to the Y input set at 0.5 V/cm. If 10 V a.c. at 50 Hz is connected to the Y input set at 5 V/cm, with the time base turned off, the spot will move so quickly up and down the screen that it will appear as a continuous line (*figure 6.4*).

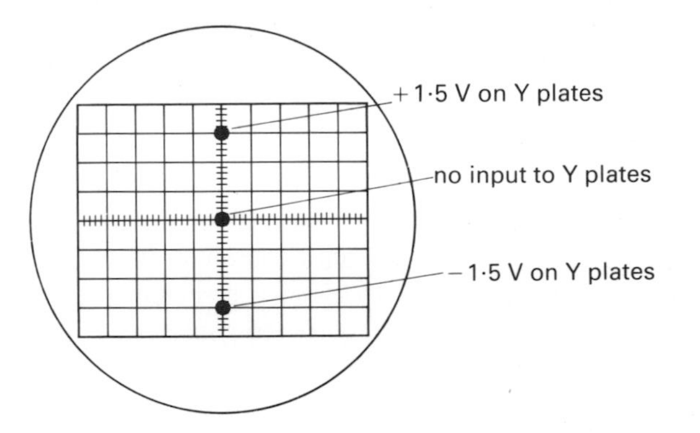

Figure 6.3 Deflection produced by d.c.

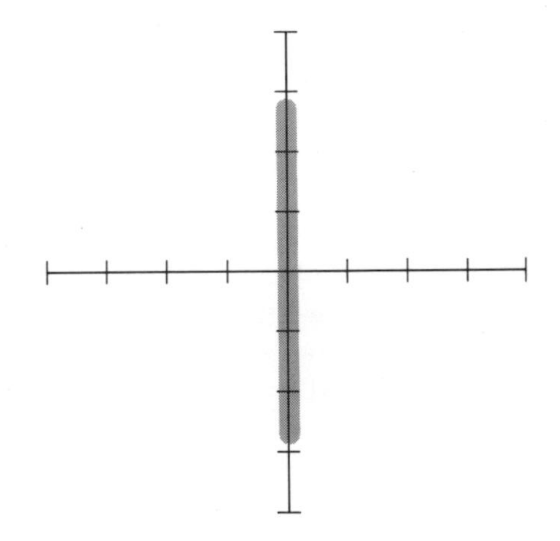

Figure 6.4 Deflection produced by a.c.

Notice that the length of the line above and below the X axis corresponds to about 15 volts, which is the peak value for the 10 V r.m.s. applied to the input. The CRO measures the peak to peak voltage of an a.c. signal, whereas an a.c. voltmeter is calibrated to read the r.m.s. value

Example A CRO produces a vertical trace 4 cm long with the Y input set at 50 mV/cm. What is the r.m.s. value?

$$\text{peak to peak voltage} = 4 \times 50\,\text{mV} = 200\,\text{mV}$$

$$\therefore \quad \text{peak voltage} = 100\,\text{mV}$$

$$\therefore \quad \text{r.m.s. voltage} = \frac{100}{\sqrt{2}} = 71\,\text{mV}$$

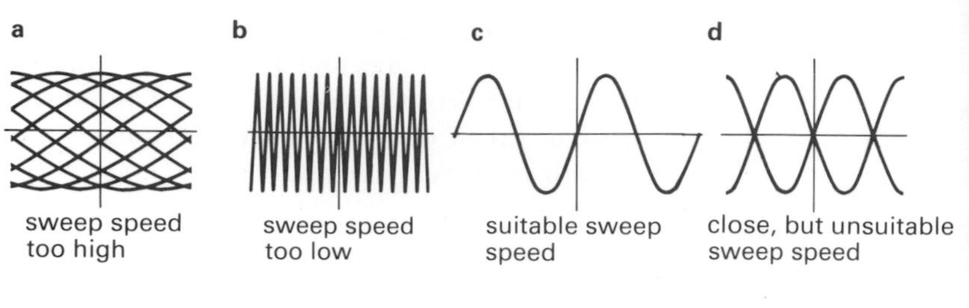

Figure 6.5

If an a.c. signal is connected to the Y input and the time base is turned on, the result should correspond to one of the diagrams in *figure 6.5*. If in *figure 6.5c* the spot takes 1 ms to cross the screen and it draws two complete waves, each wave takes 0.0005 s to complete, so in 1 s there are 1/0·0005 waves.

$$\therefore \quad \text{frequency of the signal} = \frac{1}{0{\cdot}0005} = 2000\,\text{Hz}$$

Dual beam and dual trace oscilloscopes

A dual beam CRO has two electron guns at the end of the CRT and quite independent operation of the two beams is possible, each having its own deflector plates. It is really like having two cathode ray tubes firing electrons at the same screen. The dual trace CRO is much more common; it has only one electron beam which is switched very quickly between the top half and the lower half of the screen. The time base sweeps the two traces at the same speed but two different signals (of the same frequency) can be connected to the Y_1 and Y_2 inputs. In this way we can study two signals at the same time. If the two input signals do not have the same frequency it may not be possible to freeze both of them at the same time.

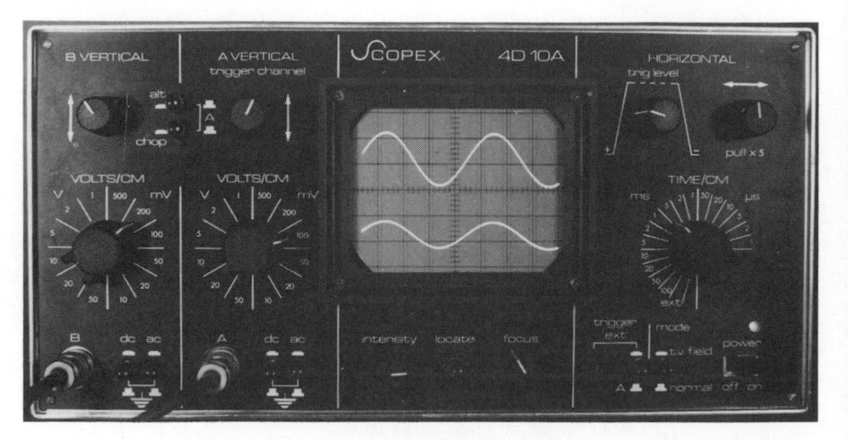

Dual trace CRO

Sources of e.m.f.

Unit 7

Primary cells

A primary cell has a source of chemical energy which it converts to electrical energy. When this energy source is exhausted, the cell has reached the end of its life. In its simplest form, it consists of two different metals surrounded by an electrolyte. An electrolyte is an electrically conducting solution and the conduction is by means of ions (unit one). A cell constructed as simply as this would, however, soon stop working owing to **polarisation**. Polarisation is the build-up of hydrogen gas around the positive electrode of the cell. In a practical cell, this hydrogen is converted to water by a chemical called the **depolariser**.

The chemical changes which occur while the cell is driving current are not designed to work in reverse and attempts to recharge such a cell are not recommended.

Primary cells are also called **dry cells**, because the electrolyte is not a liquid, but a paste. The difference between two dry cells such as SP2 and HP2 is simply that the HP2 contains more depolariser, which enables it to cope with heavier currents, such as those encountered in a portable tape recorder, without polarising. The action of the depolariser is slow and a cell which is failing may simply need time to recover. *Figure 7.1* shows the general construction of a dry cell.

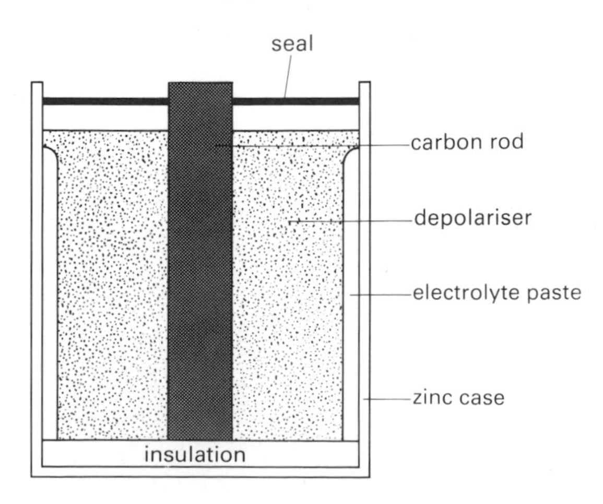

Figure 7.1 Construction of a dry cell

Secondary cells

These cells use different chemical reactions to produce the electrical energy: reactions which can be reversed by passing a current through the cells in the opposite direction to that in which they would push a current. The two most common types of secondary, or rechargeable, cell are the lead-acid type and the alkaline type. When a lead-acid cell is recharged, it produces gases and cannot, therefore, be completely sealed. The most common application of lead-acid cells is found in the car battery, where 6 of them linked together produce the required 12 V. The most popular alkaline cell is the nickel-cadmium type; when these are recharged no gas is produced, so they can be completely sealed. Nickel-cadmium cells are available in the same sizes as the most popular range of dry cells, but are much more expensive and have an e.m.f. of only 1·25 V per cell.

Secondary cells have a much lower internal resistance than dry cells, so the current which flows if their terminals are shorted together can be very heavy. A lead-acid cell is much more likely to be damaged by this than a nickel-cadmium cell.

The **capacity** (not to be confused with capacitance) of a cell is a measure of how long it can supply a current and it is measured in ampère-hours (Ah). For example, a 6 Ah cell could supply:

1 A for 6 hours
or $\frac{1}{2}$ A for 12 hours
but not 6 A for 1 hour.

Although the arithmetic is still correct in this last example, the very heavy current would cause a reduction in the capacity of the cell.

Fuel cells

These are electrochemical generators which it is hoped, before very long, will run on cheap fuel. *Figure 7.2* shows the general construction.

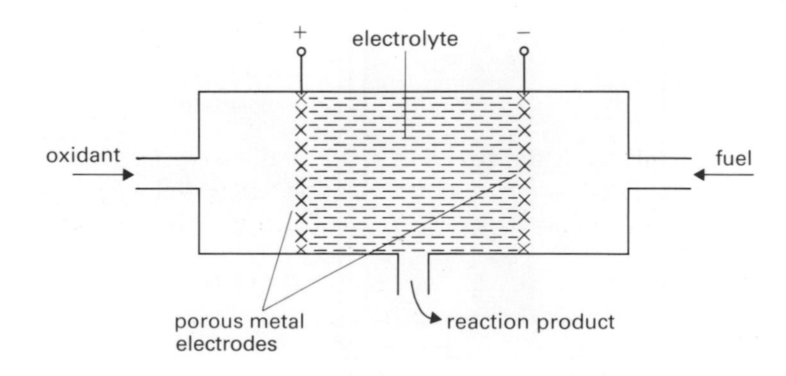

Figure 7.2

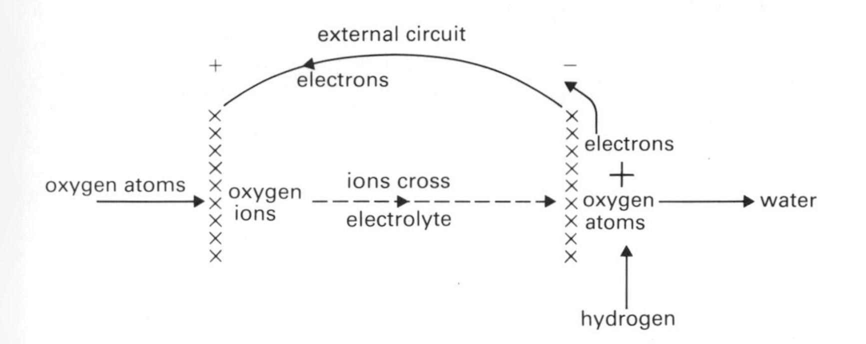

Figure 7.3 To show the action of a fuel cell

An important practical example of a fuel cell is the **hydrox cell**, as used by the Apollo spacecrafts in which the fuel was liquid hydrogen and the oxidant was liquid oxygen. The reaction product was pure water and so could be used as drinking water. The reaction is summarised in *figure 7.3*. The electrodes also act as a **catalyst** and may be constructed from nickel or platinum. The catalyst is used to speed up the chemical change without being consumed itself.

Fuel cells are uneconomical for large scale generation of electricity unless they can be designed to work on, say, natural gas and air at not too high a temperature.

Thermocouples

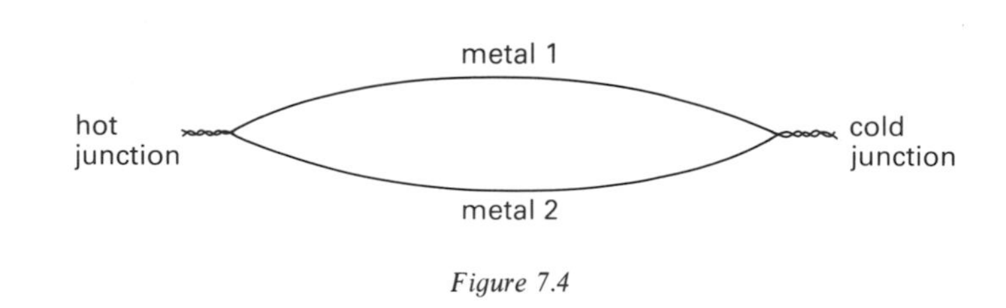

Figure 7.4

A thermocouple consists of any two metals, joined together at their ends (*figure 7.4*). If the two joins are kept at different temperatures an e.m.f. is set up. This is called a **thermoelectric e.m.f.** and its value will depend upon:

a the metals used.

b the temperatures of the two junctions.

The current which this e.m.f. will drive depends upon the total resistance of the circuit.

Thermoelectric e.m.fs. are very small, but can be measured on a sensitive meter. A platinum/platinum-iridium thermocouple has for many years formed the basis of a remote reading, high temperature thermometer (*figure 7.5*).

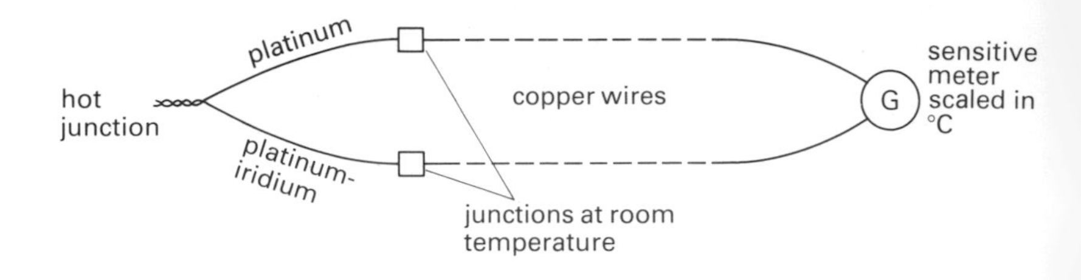

Figure 7.5
A thermocouple thermometer

The piezoelectric effect

This most unlikely method of generating an e.m.f. was first observed by the Curie brothers. They noticed that when a slice of Rochelle salt (sodium potassium tartrate) is subjected to mechanical stress, an electric charge appears across opposite faces. This happens owing to the displacement of ions within the crystal. If a thin layer of silver is deposited on these faces, connection to an external circuit is possible and the piezoelectric charge can now flow.

This is the principle of the crystal pick-up and crystal microphone, both of which use slices of Rochelle salt. Ceramic pickup cartridges use a lead zirconate ceramic bar, which produces a charge when bent. Rochelle salt gives a larger e.m.f. but it deteriorates in humid or hot conditions. The most recent application of the piezoelectric effect is based upon lead zirconate titanate. When this substance is struck, or squeezed hard, it produces a very high e.m.f. and the resulting spark can be used to ignite natural gas burners.

The reverse effect is also very useful. If a p.d. is applied to the silver electrodes, the crystal changes shape very slightly. If the applied p.d. is alternating at a suitable frequency, the crystal can be made to vibrate in a very regular and reliable way. The frequency at which this happens may be from 100 kHz to 15 MHz, according to the size and shape of the crystal. The material which is used for this purpose is **quartz**. Quartz is a natural crystalline substance; the crystals may be up to 5 cm in diameter and 10 to 15 cm long. They can be cut into thin slices and connections made to the two faces. Crystals made in this way form the basis of accurate and reliable **crystal oscillators** whose frequency varies very little with temperature. The temperature coefficient depends upon the direction in which the slice was cut from the original crystal. We use these crystal oscillators to keep radio transmitters accurately tuned to a frequency and as the basis of accurate clocks and watches.

The circuit symbol for a quartz crystal is shown in *figure 7.6* and the components to which it is equivalent in *figure 7.7*. In this equivalent circuit we can see the similarity with the simple tuned circuit, in unit three (*figure 3.17*).

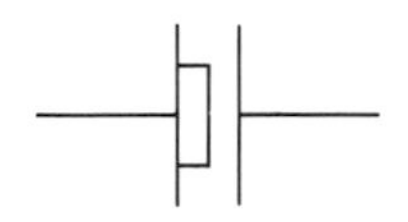

Figure 7.6
The symbol for a quartz crystal

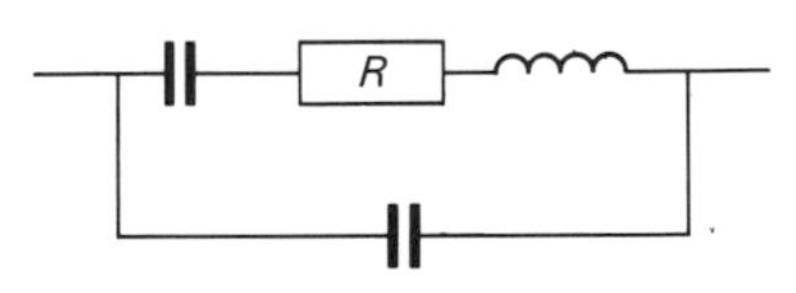

Figure 7.7
The circuit which is equivalent to a quartz crystal

Left: Quartz crystal inside a watch

Photovoltaic cells

A photovoltaic cell produces an e.m.f. when light falls upon it, so it converts light energy directly to electrical energy. The effect was first noticed in 1837, by Antoine Becquerel, when light fell on one of the electrodes in an electrolysis experiment. The effect was then noticed, in 1877, in selenium, and cells using copper (I) oxide deposited on copper have also been used. More recently, **silicon** has been used for the conversion of solar energy into electrical power. The efficiency of silicon cells is only about 10% at present. Each cell produces about 0·5 V in full sunlight with a maximum current of about 35 mA per square cm of cell.

Selenium photovoltaic cells are made by pouring the molten selenium on to a metal base plate. It is then annealed at a temperature just below its melting point, to convert the selenium into a crystalline semiconducting form. An insulating barrier layer forms on the surface during annealing and a transparent layer of gold is evaporated on to this barrier layer. Electrodes are then connected to the metal base and the gold layer (*figure 7.8*). The metal base becomes positive and the gold layer negative when light shines on the cell through the gold layer. The efficiency is only about 0·5%, so these cells are used for photographic light measuring meters, rather than for power generation.

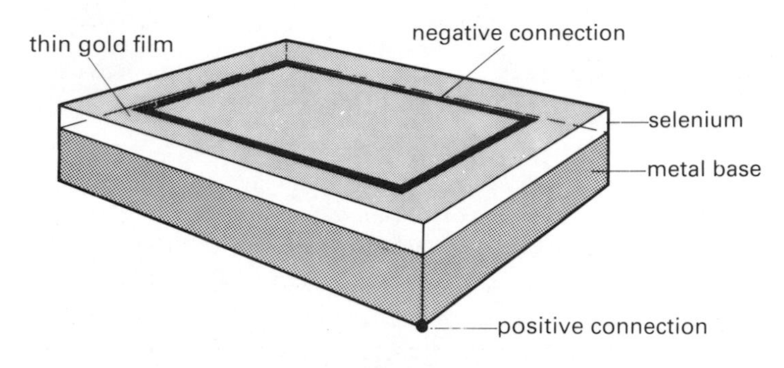

Figure 7.8 A selenium photovoltaic cell

Semiconducting junctions may also be used as photovoltaic cells, or, when reverse biased (see unit twelve), as photoconducting devices. Notice the difference between **photovoltaic**, meaning the generation of an e.m.f. by the action of light, and **photoconducting**, meaning a device whose resistance varies according to the intensity of light energy falling on it.

Electric lamps

Filament lamps

Filament lamps are also known as incandescent lamps; they give out light because an electric current is passing through a thin, high resistance filament and raising its temperature until it becomes white hot. The most commonly used filament material is tungsten metal, which has a melting point of 3390°C or 3663 K (Kelvin). A normal domestic lamp operates with a filament temperature of about 2600 K and the extra-bright, short lived photoflood lamp used by photographers may have a filament temperature of 3200 K. Since the colour of the light given out is determined by the filament temperature, it is usual to describe any light similar in colour to, say, a photoflood lamp as having a **colour temperature** of 3200 K.

Tungsten–argon filament lamps

If the tungsten filament is surrounded by a vacuum, there is a tendency for the tungsten to evaporate and condense on the cool glass, thus darkening it. If the glass envelope is filled with argon gas at low pressure, then when the lamp is running, the argon pressure rises to about the same as that of the atmosphere outside. Under these conditions, there is less evaporation from the filament, but heat is now carried by the gas from the filament to the glass. This loss of heat from the filament means a lowering of efficiency of the lamp, and the glass becomes much hotter than with a vacuum lamp. This loss of efficiency can be reduced by using a **coiled coil** filament as shown in the photograph below.

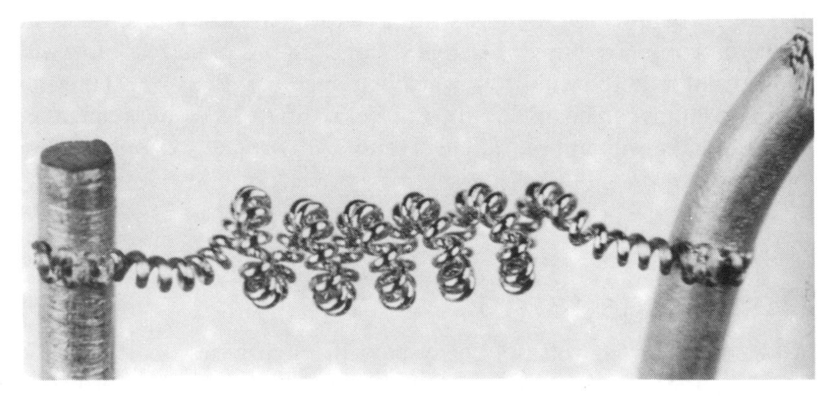

Coiled coil lamp filament

It is worth noting that a filament lamp should always be operated at its correct voltage, because a 2% rise above the correct voltage shortens the life of the lamp by about 10%.

Tungsten halogen lamps

If the tungsten filament is surrounded by iodine vapour and the envelope is made of quartz, a much higher filament temperature can be used. Under these conditions, the tungsten vapour leaving the filament combines with the iodine to make the colourless tungsten iodide. This cannot darken the glass and when it touches the hot filament, the tungsten and iodine are regenerated. The filament is, therefore, being built up and evaporated continuously, which prolongs its life at this higher temperature. A high operating temperature is essential for these lamps, and the quartz envelope must not be touched with fingers. More light, of bluer colour, is produced for the same power as a conventional filament lamp. Because they contain quartz and iodine, they are also called **quartz iodine** lamps.

Carbon filament lamps

Joseph Swan demonstrated his first carbon filament lamp in 1879, but they were replaced by tungsten filaments in 1911. However, there has been a renewed interest in carbon filament lamps for decorative purposes, and modern methods of making the filament give a reliable lamp. It operates at a lower temperature than tungsten and gives a softer, yellower light.

Reflector lamps

Filament lamps with built-in reflectors are commonly met as sealed beam headlamps for cars. The most interesting type of reflector lamp is the dichroic type used in small gauge film projectors. These reflect the light, but not the heat, towards the film.

Working voltage and power of filament lamps

If a small lamp is required, it will have to be a low voltage one. If a powerful lamp is required, a mains voltage lamp is preferred, otherwise the current required will be very heavy (remember, $W = VI$). Tungsten filament lamps can be made with a power of up to 3 kW, but remember, most of this will appear as heat and can present serious cooling problems.

Discharge lamps

An electric current will not easily pass through gases or vapours. If, however, the p.d. is high enough to ionise the gas, the ions will provide a good conducting path for the current. As the ions re-combine to form the neutral gas atoms, light of characteristic colour is emitted. The most common examples of discharge lamps are:

Low power neon lamps for use above 100 V. They are used very often as mains indicator lamps.
Mercury vapour (blue-green), sodium vapour (orange-yellow) and high pressure helium (pale pink), for street lighting.
Low pressure mercury vapour in fluorescent tubes.
High voltage, inert gases (helium, neon, argon, krypton, xenon etc.) in coloured advertising displays.
Xenon in photographic flash tubes.

Discharge lamps are more efficient at converting electrical energy into light energy, but they still produce heat.

The electroluminescent panel

All filament and gas discharge lamps give out a large proportion of the electrical energy they convert not as light, but as heat. The nearest thing to a cold light source is a device called an electroluminescent panel. They are not often used because they produce green light, and not much of it: to illuminate a room with one, it would have to cover the entire ceiling and the cost would be prohibitive. They have been sold as nursery night lights, but their eerie glow is not very comforting.

Light-emitting diodes

Semiconducting diodes (see unit twelve) always emit some kind of radiation when they are forward biased. The normal silicon diode produces no useful radiation, but gallium arsenide and gallium phosphide diodes emit light. The colour of the light will depend upon the material used for the diode and at present red, yellow and green light emitting diodes are available.

The light-emitting diode (LED for short) forms the basis of the numerical display in many electronic calculators. It can produce light with almost no heating of the diode. The lowest current filament lamp requires about 40 mA while an LED operates with a current of about 5 mA.

There are two important precautions to be observed when using an LED:

a The LED must be connected with the cathode made negative and the anode positive.

b A series resistor must always be used to limit the current and its value R is calculated using the equation:

$$R = \frac{\text{supply voltage} - 2}{0{\cdot}01}$$

The circuit symbol for an LED is shown in *figure 8.1*. When looking at an LED there is no general rule for knowing which lead is the cathode and which the anode. An ohm-meter used as shown on page 96 will identify the leads of a diode, including an LED.

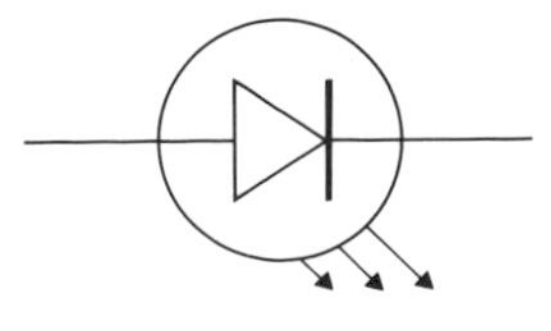

Figure 8.1
Circuit symbol of a light emitting diode

Unit 9

Transducers

A transducer is a device which converts one type of energy into another. This unit is concerned with electrical transducers, that is, those in which electrical energy is either the input or the output.

d.c. electric motors

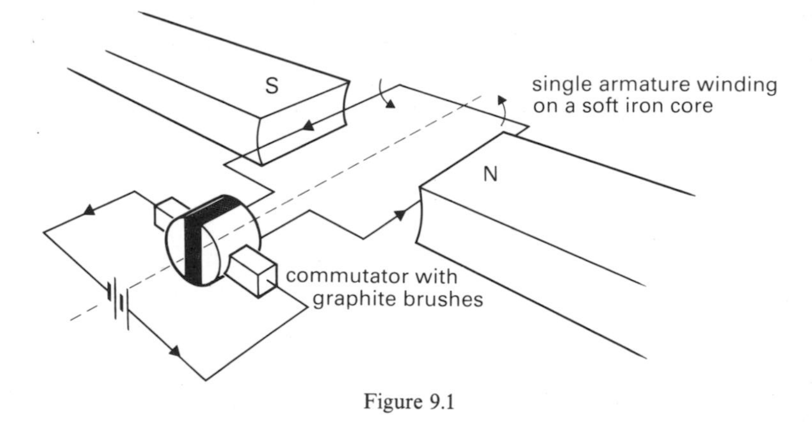

Figure 9.1

Figure 9.1 shows the diagram of a simple d.c. two-pole motor. It is called two-pole because the rotating armature has one north and one south pole. This motor is not self starting, unless by chance it stops in the position shown. It is in this position that the maximum turning force or **torque** is exerted on the armature. A motor with three or more poles on the armature will be self starting.

The main use for this type of motor is for working small models. More powerful motors, for example in electric trains, would have the permanent magnet replaced by an electromagnet (called the field winding). The armature and field windings may be connected in series or in parallel (shunt). These different methods of connection give rise to different properties:

A series wound motor has a high starting torque and can be connected to the full supply voltage, even when at rest. The objections to series wound motors are that the speed varies with the mechanical load and the motor races when the load is removed.

A shunt wound motor runs at almost constant speed with varying load and would be suitable for use on a lathe or drilling machine.

A compound wound motor has coils connected as shown in *figure 9.2*; it has the combined advantages of almost constant speed with a high starting torque.

All large electric motors require a starting resistor in series with the supply, to limit the current through the windings until the motor has built up speed. As soon as the motor begins to move, a back e.m.f. is set up which opposes the current flowing in the armature (remember Lenz's law, unit 3).

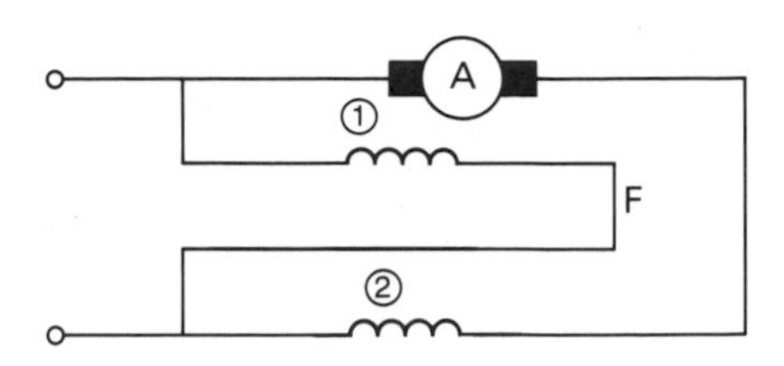

A = armature coil
F = field coils with coil 1 in parallel and coil in series 2

Figure 9.2
A compound wound motor

Induction motors

These motors will work on a.c. only and may be of two possible types, large industrial **squirrel cage motors** and small **shaded pole motors** for record players, mains tape recorders, and electric clocks. The words 'squirrel cage' refer to the open basket-work type of construction of the rotor. The words 'shaded pole' refer to the action of a piece of copper in those motors which 'shades' the magnetic field from the rotor.
The advantages of induction motors are:

- **a** constant speed controlled by the mains frequency and not the supply voltage.
- **b** no brushes to wear and no sparks to cause radio interference.
- **c** silent operation.
- **d** low power consumption and no starting resistor needed.

The disadvantages of the induction motor are:

- **a** relatively low power for their size.
- **b** low starting torque.

Generators

A simple d.c. generator would have the same construction as the simple motor in *figure 9.1*. An e.m.f. is generated when a conductor cuts through a magnetic field. When the armature is made to turn, the e.m.f. generated varies as shown in *figure 9.3*. Although the output is d.c., it is not smooth d.c. and it might be better to call it uni-directional. Smoother d.c. would be obtained by having many separate coils on the armature and a corresponding number of contacts on the commutator. Such generators are seldom used.

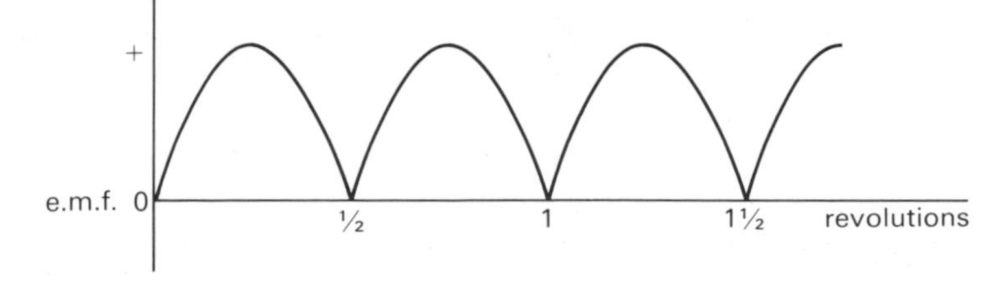

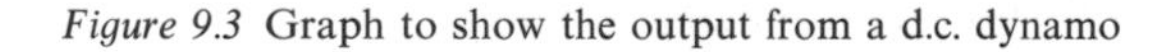
Figure 9.3 Graph to show the output from a d.c. dynamo

A car dynamo is a d.c. generator with a multipole armature and field coils, instead of a permanent magnetic field. The current generated by the armature provides the current to energise the field coils, and there is always sufficient residual magnetism in the soft iron cores to start this process. Cars are now being fitted with a.c. dynamos or **alternators**, because these produce a higher output at low engine speeds. The a.c. they produce is rectified (changed to d.c.) and fed to the field coils and out via slip rings to charge the battery.

In power stations, the output of a single alternator may be at a pressure of 25 kV with a current of several hundred amps. This cannot be led out of the alternator via slip rings because the sliding contacts would soon be destroyed by the sparking. The fixed coils or stator (*figure 9.4*) are the coils in which the e.m.f. is set up; the output from these can be taken via fixed wires. The rotating coils (rotor) provide the magnetic field, and the relatively small current required to energise the rotor can be fed in through slip rings. The d.c. for this purpose is generated by an **exciter** driven by the same shaft as the rotor.

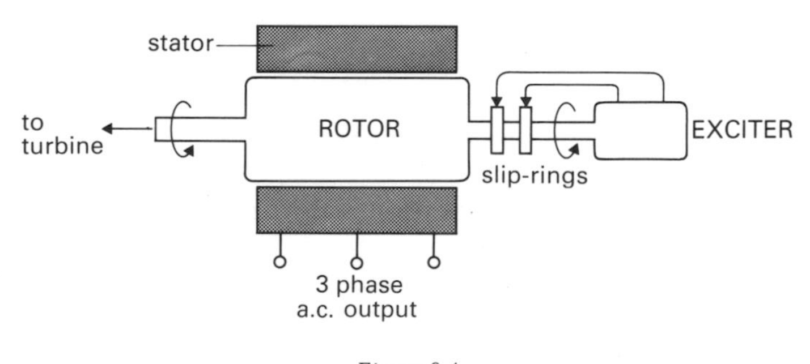

Figure 9.4
A power station alternator

The bicycle dynamo is an unusual a.c. dynamo as it has fixed output coils and a rotating permanent magnet.

Microphones

There are basically two types of microphone, namely **constant velocity** and **constant amplitude**.

In a constant velocity type the output e.m.f. is proportional to the **velocity** of vibration of the moving parts, and it is independent of the amplitude or amount. The **moving coil** or dynamic (*figure 9.6a*) and **ribbon microphone** (*figure 9.6b*) are examples of this type. The e.m.f. in these two microphones is generated as the moving conductor (the coil or the ribbon) cuts through the magnetic field.

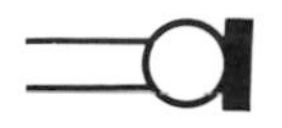

Figure 9.5
Circuit symbol for a microphone

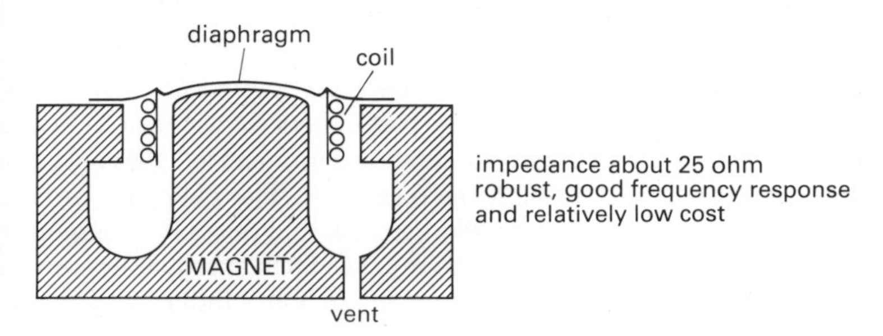

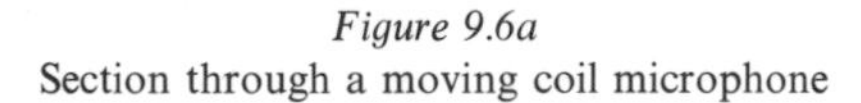
Figure 9.6a
Section through a moving coil microphone

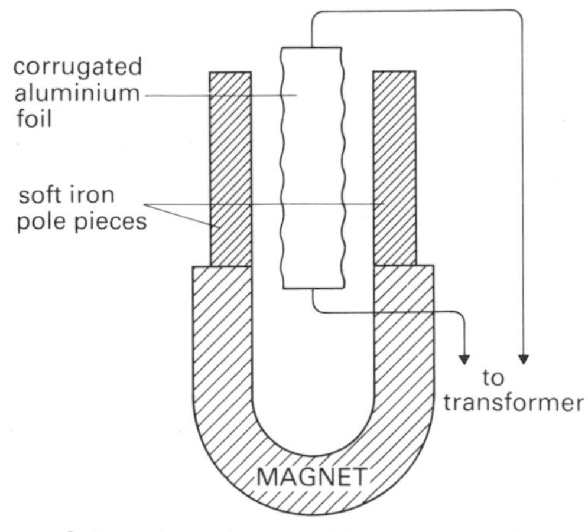

Quite robust, but sensitive to blasts of air

Figure 9.6b
Simplified diagram of a ribbon microphone

In a constant amplitude microphone, the output is proportional to the amount by which the sound waves **displace** the moving element from its rest position. The **crystal** and **capacitor** microphone are examples of this type. The e.m.f. in the crystal microphone is set up by the piezoelectric effect (see unit seven). The structure of the crystal (*figure 9.7*) requires some explanation. If a single crystal is bent, one face is stretched and the opposite face is compressed. If the connecting wires were simply connected to these faces, the piezoelectric e.m.fs. would tend to cancel by these opposite movements. If the crystal is made in two parts, as shown, the e.m.fs. can be made to add. Such a crystal is called a **bimorph**.

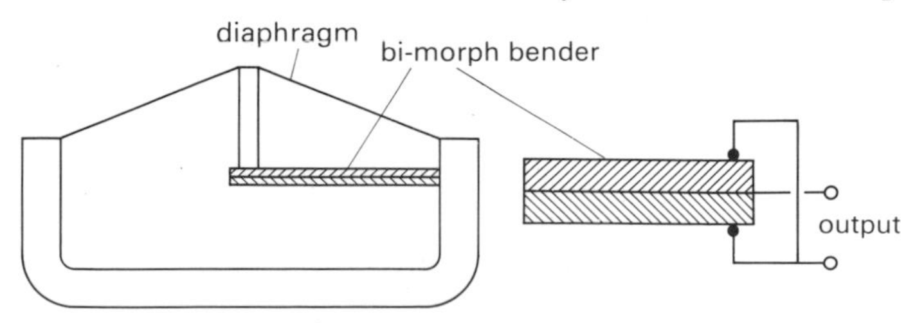

Has a very high impedance – usually over 1 MΩ
High output and very cheap

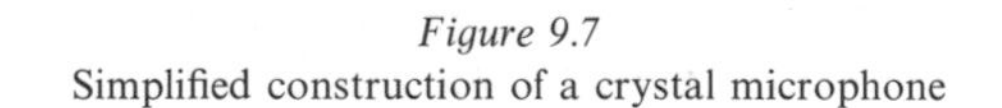
Figure 9.7
Simplified construction of a crystal microphone

The capacitor microphone element generates no e.m.f. at all; it simply varies its capacitance as the moving plate alters its distance from the fixed plate. By applying a small p.d., as shown in *figure 9.8*, the capacitor charges and discharges through the resistor R. This changing current causes a small varying p.d. to appear across R and this is then amplified before being sent along the microphone cable. Such microphones can be powered by a small 1.5 V cell contained in the body; the cell is switched on and off as required and will have a life of over 1000 hours, as it supplies so little current.

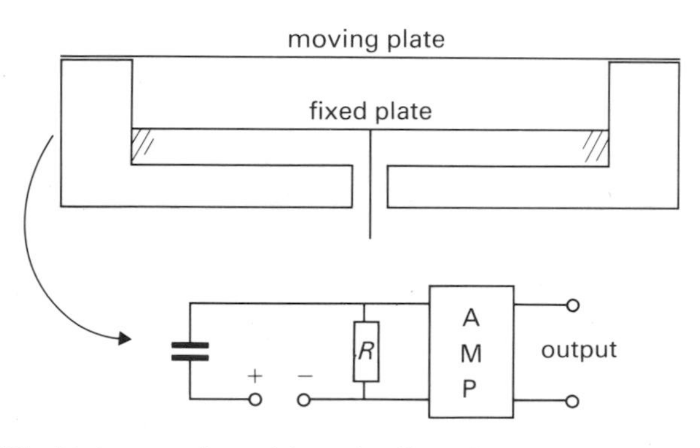

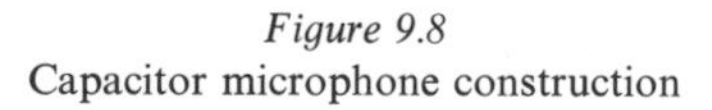

Figure 9.8
Capacitor microphone construction

All of these microphones are capable of producing a reasonably high quality output. The **carbon microphone** (*figure 9.9*), as used by the post office in their telephones, is not really in the same class. The main reason for its continued use is the presence of about 50 V d.c. in the telephone wires. This voltage is necessary for the operation of the carbon microphone and a hindrance for other types.

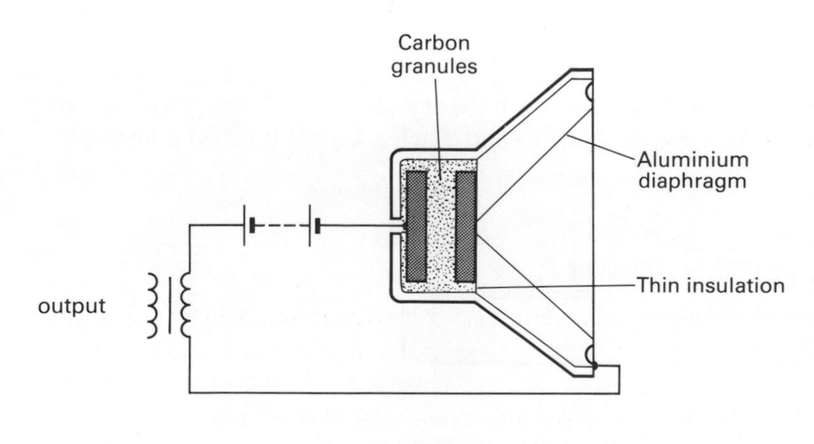

Figure 9.9
Section through a carbon microphone

Directional properties of microphones

A microphone is more sensitive to sound waves approaching it from certain directions and less sensitive in other directions. In order to describe these directional properties in a pictorial way, we make use of special polar graph paper (*figure 9.10*). The lines radiating outwards from the centre are used to show the direction of the incoming sound. The distance of a point along one of these lines is a measure of the microphone sensitivity. *Figure 9.10* shows the polar graph for a ribbon microphone; as it is sensitive in two different directions, it is said to be bi-directional or 'figure of eight'.

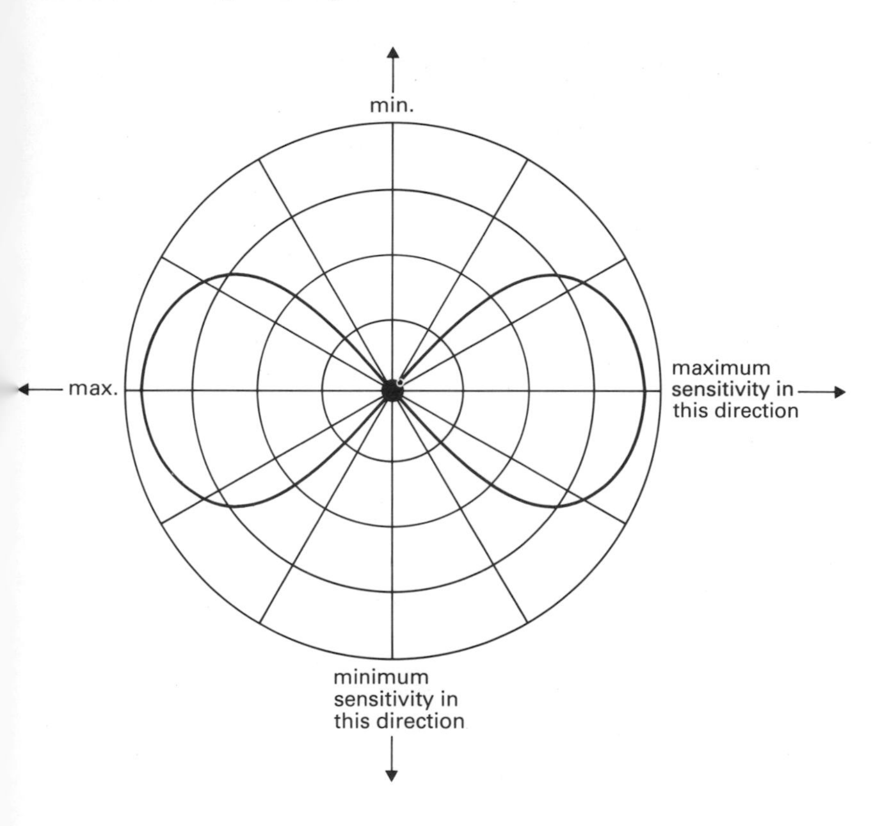

Figure 9.10 Polar graph of microphone sensitivity – the microphone is at the centre of the diagram.

The microphone cable

If a microphone has an impedance of less than about 600 Ω it is possible to have a very long cable between the microphone and the amplifier without the cable picking up unwanted **hum**. If the microphone is a high impedance crystal type, only short connecting cables can be used. It is also important to use special microphone cable if unwanted noises due to the movement of the cable are to be avoided.

Loudspeakers

The majority of loudspeakers in use are of the moving coil type shown in *figure 9.11*. When a current flows through the voice coil, the magnetic field it produces interacts with the loudspeaker's permanent magnet. The coil is then pulled in or pushed out, according to the direction of the current. The cone, which is firmly attached to the coil, passes this movement on to a large area of air. These pressure changes in the air correspond to the sound wave which produced the electrical signal in the first place.

As the construction of a moving coil loudspeaker and a moving coil microphone are so similar, it is not surprising that a loudspeaker can act as a low impedance microphone, and indeed does, in a simple two-way intercom.

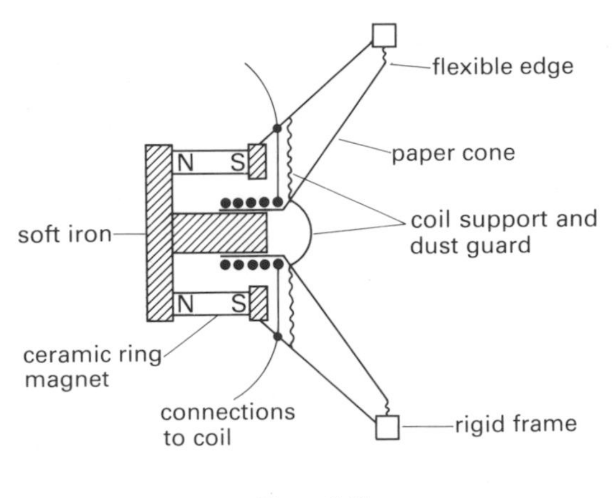

Figure 9.11
Section through a moving coil loudspeaker

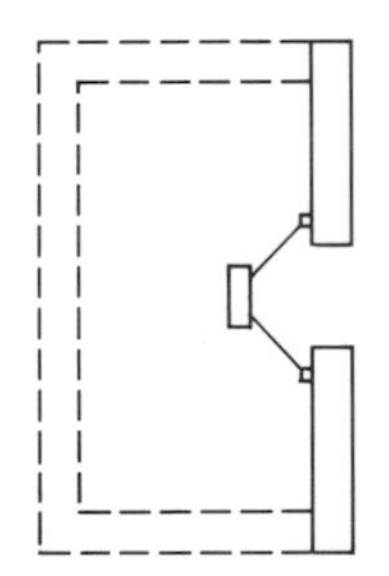

Figure 9.12
A loudspeaker baffle

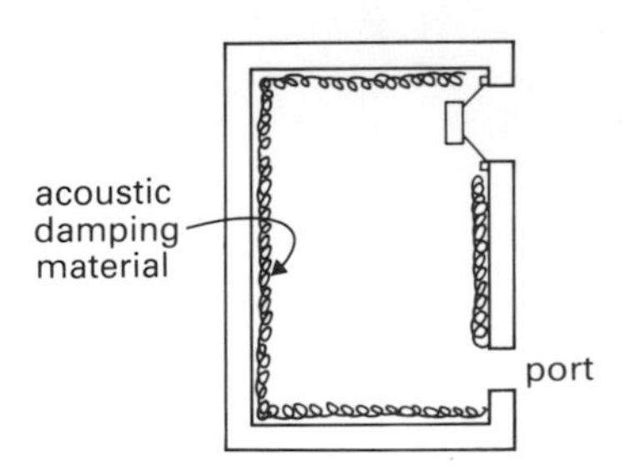

Figure 9.13
Loudspeaker reflex enclosure

Baffles and enclosures

Sound waves are radiated from both the front and the back of a loudspeaker cone. If the cone moving forwards produces the 'crest' of the wave, then the back of the cone will produce a 'trough' at the same time. If the crest and the trough are allowed to meet, they will partly cancel. The earliest solution to this problem was to use a large flat **baffle** to screen off the waves from the back of the speaker (*figure* 9.12). The effectiveness of this arrangement could obviously be improved by closing in the sides and back, as shown by the dotted lines in *figure 9.12*. This was called the **infinite baffle** and it certainly prevented the waves from the back meeting the waves from the front. However, this totally enclosed volume of air restricted the movement of the cone, since it had to be compressed as the cone moved back into it. By making a suitable size hole, or **port**, in the correct place, this stiffness can be reduced, and the sound emerging from the port can be arranged to be **in step** with the sound direct from the cone. This type of loudspeaker enclosure is called a **bass reflex** cabinet (*figure 9.13*).

Loudspeakers which are designed to radiate sound at around 25 Hz tend not to be very efficient radiators at around 16 kHz. It is, therefore, quite usual to find in a loudspeaker cabinet of any quality, at least two loudspeakers. The one designed to radiate low frequencies is called the bass unit or **woofer**, and the one designed for the high frequencies is called the treble unit or **tweeter**. There may also be a mid-range unit. In order to supply each loudspeaker with its correct range of frequencies a two-way or three-way **crossover** unit is required. Two circuits for two-way crossover units are shown in *figure 9.14*. Think about the reactance of capacitors and inductors at high and low frequencies and see if you can follow:

a the path of high frequencies to the tweeter only.
b the path of low frequencies to the woofer only.

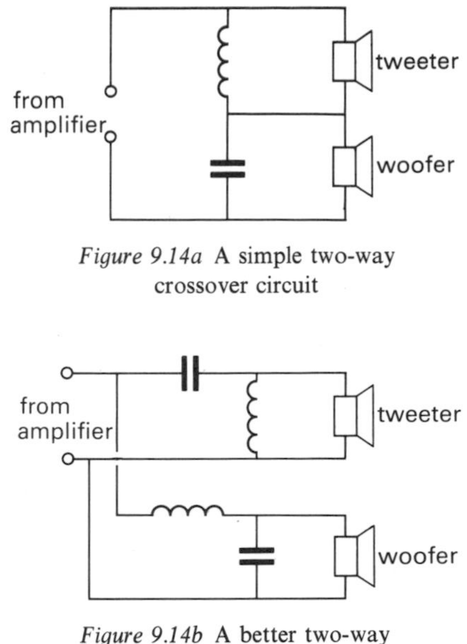

Figure 9.14a A simple two-way crossover circuit

Figure 9.14b A better two-way crossover circuit

Elliptical loudspeakers

Loudspeakers with elliptical cones are frequently met in television sets and record players. The design is something of a compromise between a loudspeaker which will radiate reasonably well at bass frequencies and one which will not take up too much space.

Electrostatic loudspeakers

The electrostatic loudspeaker manufactured by Quad has long been recognised for its natural sound. The diaphragm on electrostatic speakers moves to and fro much less than the cone of a moving coil loudspeaker. This means they have to have a large area in order to move sufficient air. Very high voltages are required for the operation of these loudspeakers, and the necessary components for generating them are built into the loudspeaker enclosure. Two miniature electrostatic loudspeakers may be used as headphones and in this case the high voltage power supply is contained in a separate box.

Power handling capacity

The maximum power a loudspeaker can handle is normally quoted as the r.m.s. power in watts. If it is driven at powers above this, the cone will be made to move more than was intended and distortion of the sound will result. In cases of exceptional overload, permanent damage may occur to the coil or the cone. Loudspeakers are not generally considered to be repairable.

Using stereo loudspeakers

When using two loudspeakers in a stereo system, the correct position for the listener is that of one of the vertices of an equilateral triangle, with the loudspeakers forming the other two vertices. Ideally, this triangle should have sides of minimum length 2 m and maximum length 4 m. The loudspeakers should be angled towards the listener.

Phasing of loudspeakers

Whenever two or more loudspeakers are in use at the same time, it is necessary for all the cones to move forwards and backwards together. If this does not happen, the sound waves from one speaker will tend to cancel with the waves from the other. The result of the left and right loudspeakers being out of phase is the production of sounds which do not appear to have a fixed position in space. This effect is not easy to detect and the best solution is to play a stereo test record such as E.M.I.'s 'The Enjoyment of Stereo'. The cause of this condition is often found to be a reversal of the connections to one of the loudspeakers.

Voice coil impedance

As the voice coil of a loudspeaker has inductance, its opposition or impedance to the alternating currents through it will vary with frequency (unit three). The combined effect of the resistance of the coil wire and its reactance is called the **impedance** (Z)

$$Z = \sqrt{R^2 + X_L^2}$$

where R = resistance of coil in ohms.

X_L = inductive reactance in ohms.

Z = impedance in ohms.

As the impedance increases with increasing frequency, it is usual to quote its value at 1 kHz, typical values being 4 Ω, 8 Ω or 16 Ω

At a frequency of 1 kHz, the value of X_L is quite small and the use of an Avometer to measure the d.c. resistance gives a value close enough to the impedance for most practical purposes. Thus, a loudspeaker of impedance 4 Ω will have a resistance of about 4 Ω when measured with an ohmmeter.

The control of current

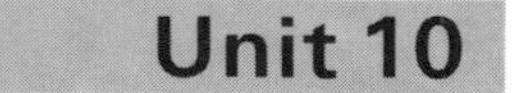

Control of current in d.c. circuits

Resistors are used in series in d.c. circuits for the control of current. If some adjustment is necessary, a variable resistor or **rheostat** is used. Remember, it is necessary to calculate the power dissipated by the resistor, and to select a suitable power rating for the component (unit two). A rheostat may be connected in two ways as shown in *figure 10.1*. In the first diagram, the circuit would be broken if the slider lost contact, but in the second the resistor remains in circuit, even if slider contact is lost.

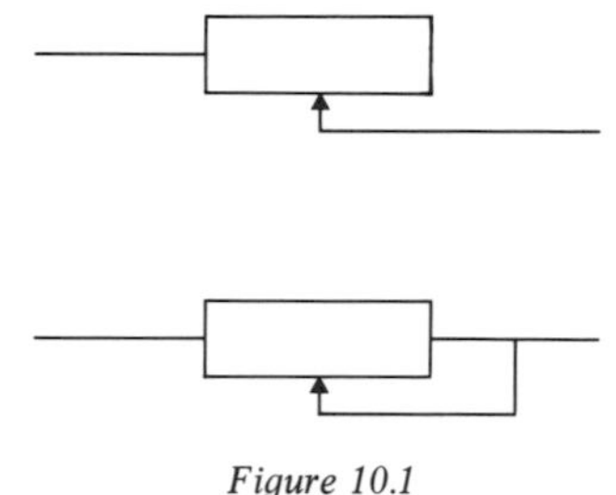

Figure 10.1
Circuit symbols for variable resistors

For very fine control over a wide range of currents, use one of the circuits shown in *figure 10.2*.

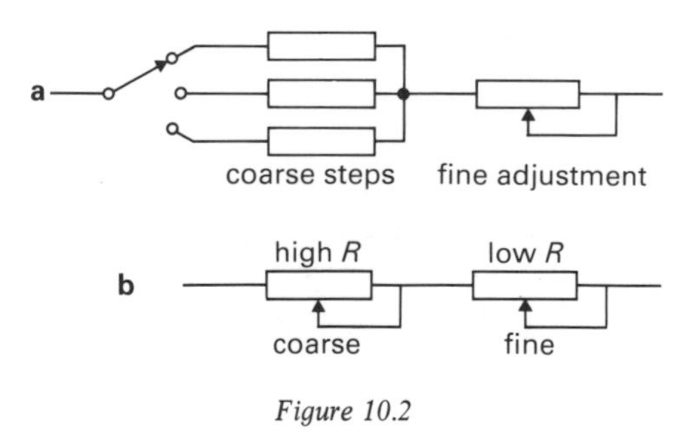

Figure 10.2

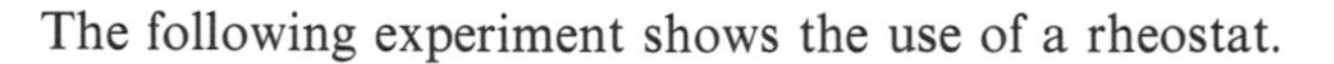

The following experiment shows the use of a rheostat.

Design a circuit to check the 1 mA calibration mark of one meter with the 1 mA mark on another, using a 2.6 V nickel-cadmium battery.

If the circuit in *figure 10.3* is used, the same current must pass through both meters, and the aim is to be able to make fine adjustments close to 1 mA. Calculate the value of R:

$$V = IR$$

$$2{\cdot}6 = 0{\cdot}001 \times R$$

$$\therefore \quad R = 2600\,\Omega$$

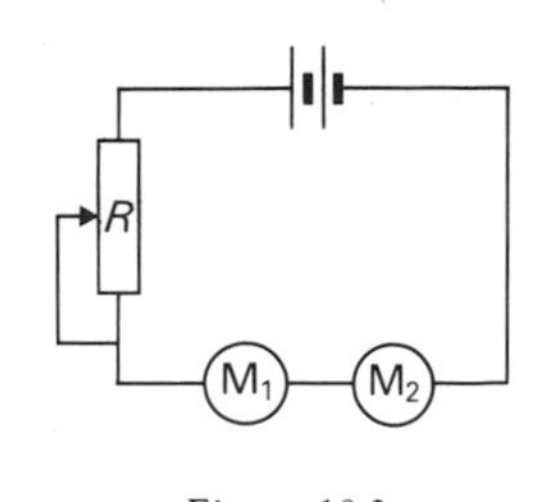

Figure 10.3

The nearest value of rheostat likely to be found is 5 kΩ, and this would not give a very fine control. A better arrangement is shown in *figure 10.4*, with the fixed resistor setting the current just above 1 mA and the rheostat giving the final adjustment.

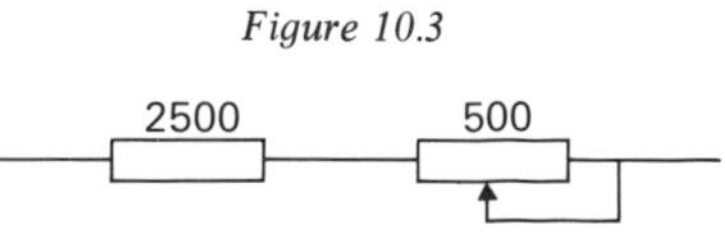

Figure 10.4

Control of current in a.c. circuits

A resistor can be used to control current in an a.c. circuit and the value required can be calculated by applying Ohm's law to the r.m.s. values of current and p.d. Because, however, heat is always produced by a current flowing through a resistor, it is preferable to use an inductor to control current in a.c. circuits. Except under the rare conditions of superconductivity (p. 78), an inductor has both resistance and reactance. Both of these exercise control over a.c. and the combined opposition is called the **impedance** of the inductor. However, the resistance and reactance cannot simply be added together, and the following equation must be used:

$$Z = \sqrt{R^2 + X_L^2}$$

where R = resistance in ohms
X_L = inductive reactance in ohms
Z = impedance in ohms
$X_L = 2\pi f L$
f = a.c. frequency in Hz.

To calculate the current flowing, use Ohm's law thus: $I = \dfrac{V}{Z}$

Where V = r.m.s. voltage
Z = impedance in ohms

Example A 1 H inductor of resistance 100 Ω is connected across **a** 250 V d.c. and **b** 250 V 50 Hz a.c. Calculate the current flowing in each case.

a

$$I = \frac{V}{R}$$

$$I = \frac{250}{100}$$

$$= 2{\cdot}5\ \text{A}$$

b

$$X_L = 2\pi f L$$

$$X_L = 2 \times 3{\cdot}142 \times 50 \times 1$$

$$\therefore\ X_L = 314\ \Omega$$

$$Z = \sqrt{R^2 + X_L^2}$$

$$= \sqrt{(100)^2 + (314)^2}$$

$$= \sqrt{108\,700}$$

$$\therefore\ Z = 330\ \Omega$$

$$I = \frac{V}{Z}$$

$$= \frac{250}{330}$$

$$= 0{\cdot}76\ \text{A}$$

Heat will be produced in both cases. The rate at which they produce heat is now calculated using: Heat = I^2R

in part **a** Heat produced $= (2{\cdot}5)^2 \times 100$

$= 6{\cdot}25 \times 100$

$= 625$ J/S

in part **b** Heat produced $= (0{\cdot}76)^2 \times 100$

$= 58$ J/s

If a capacitor is being used to control a.c. then only a little heat is produced. We can again apply Ohm's law as follows:

$$I = \frac{V}{X_c}$$

where: V = applied voltage

X_c = capacitative reactance $= \dfrac{1}{2\pi f C}$

Capacitors are generally only used in this way for controlling high frequency a.c.; at low frequencies the value of capacitance needed to give a usefully low reactance would be rather large.

Example A capacitor is to be used to operate a 27 V 0·3 A lamp from a 240 V a.c. mains of frequency 50 Hz. Calculate the value of capacitor required and its working voltage. *Figure 10.5* shows the circuit.

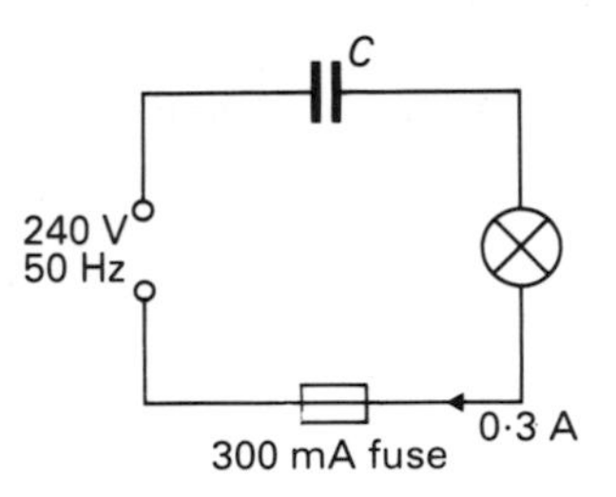

Figure 10.5
Using a capacitor to control a.c.

$$I = \frac{V}{X_c} \qquad \text{(ignore lamp resistance)}$$

$$0{\cdot}3 = \frac{240}{X_c}$$

$$X_c = 800\,\Omega$$

$$\therefore \quad \frac{1}{2\pi f C} = 800\,\Omega \qquad \text{if } C \text{ is in F}$$

$$\frac{10^6}{2\pi f C} = 800\,\Omega \qquad \text{if } C \text{ is in } \mu\text{F}$$

$$\therefore \quad C = \frac{10^6}{2\pi \times 50 \times 800}$$

$$= \frac{100}{8\pi}$$

$$\therefore \quad C = 4\mu\text{F (almost)}$$

The working voltage would have to be the **peak** value of the a.c.

peak voltage $= 1{\cdot}414 \times$ r.m.s. voltage

$= 1{\cdot}414 \times 240$

$= 339$ V

Thus a 4 μF 350 V capacitor could be used. This capacitor could **not** be an electrolytic one; can you see why? Why do you think a 300 mA fuse is included in the circuit?

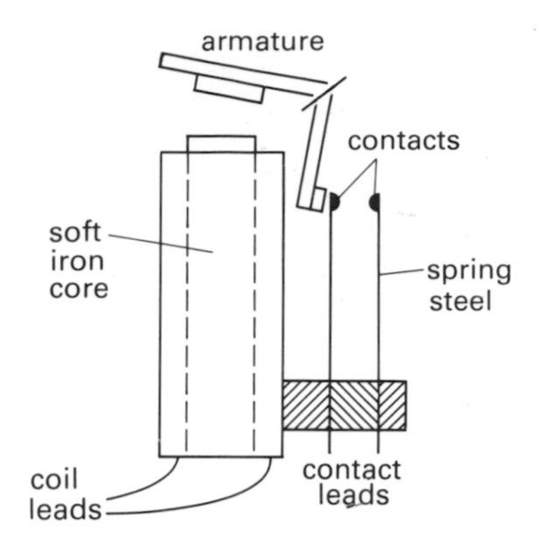

Figure 10.6
A post office type relay

Relays and reed switches

Resistors, capacitors and inductors can be used to alter the strength of current flowing in a circuit. If you wish to turn that current on and off, you use some form of switch. If a simple, hand operated switch is not suitable, then perhaps a **relay** could be used. *Figure 10.6* shows a **post office** type relay, so called because of their use in telephone exchanges. By passing a small current through the coil, a magnetic field is produced and the armature is attracted to the core. This movement of the armature can then be used

a for switching large currents.

b for simultaneous operation of many different switch contacts.

A relay switch may be made up from any combination of **make**, **break** and **change-over contacts** (*figure 10.7*).

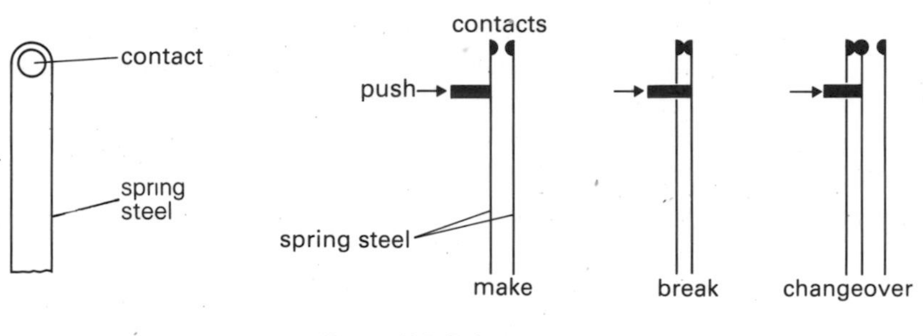

Figure 10.7 Relay contacts

The relay coil would be placed in series with a circuit if operation by a current was required; it would be placed across the supply if voltage operation was required.

The post office relay is relatively slow in operation and is most suitable where the operation of a single switch in one place makes several switches work somewhere else. If a faster, single switching operation is required, a reed relay would be used (*figure 10.8*). If a strong magnetic field is brought close to the reed the contacts meet or change over. This field may be provided by a nearby permanent magnet, forming a **reed switch** or by a current flowing in a coil around the reed, forming a **reed relay**.

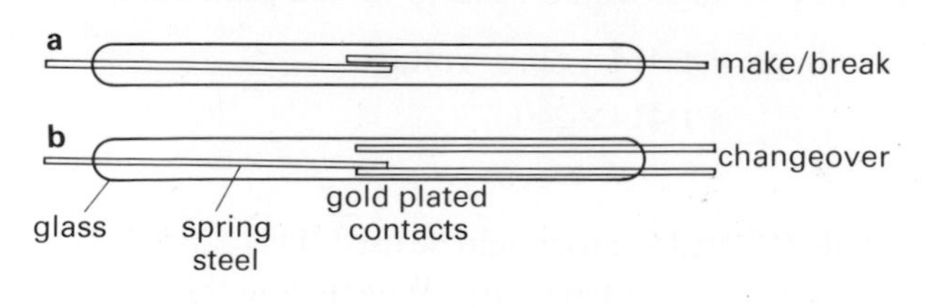

Figure 10.8 Reed switches

The back e.m.f. of a relay

If the current through a relay coil suddenly falls to zero, quite a large back e.m.f. is set up. This voltage could damage the transistor controlling the current through the coil. If a diode is connected across the coil as shown in *figure 10.9* the back e.m.f. flows through the diode and is prevented from building up to a high value likely to damage the transistor.

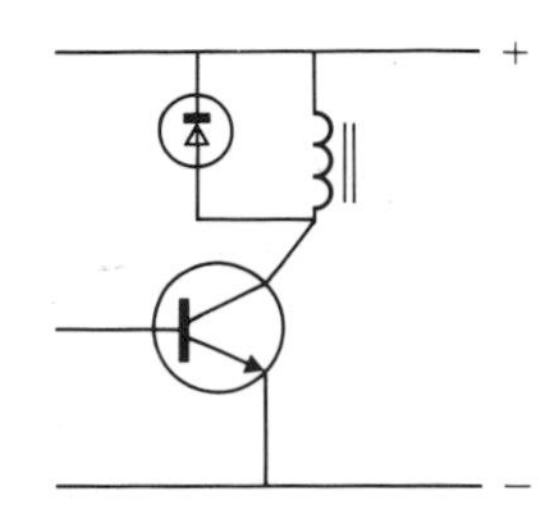

Figure 10.9 Protecting a transistor from back e.m.f.

The potential divider or 'pot'

The use of two resistors to act as a potential divider has already been mentioned in unit four. If the two fixed resistors are replaced by a single variable resistor, as in *figure 10.10*, continuous adjustment of the output between zero and V volts is possible. The actual value chosen for the variable resistor is not too critical, but it must be high enough to permit only a few milliamps to be drawn from the supply.

This device is frequently, but not entirely correctly, known as a potentiometer. The name potential divider is preferred because, provided only tiny currents are being drawn from the output terminals, the output voltage is proportional to the rotation of the control.

In a **linear** potential divider, the resistance of the track from the zero end increases in proportion to the degree of rotation. In a **log pot** the resistance varies logarithmically with the rotation; such a pot would be used as a volume control in an amplifier, so that the volume increases in proportion to the rotation. Other variations such as semi-log and inverse log are also available for special purposes.

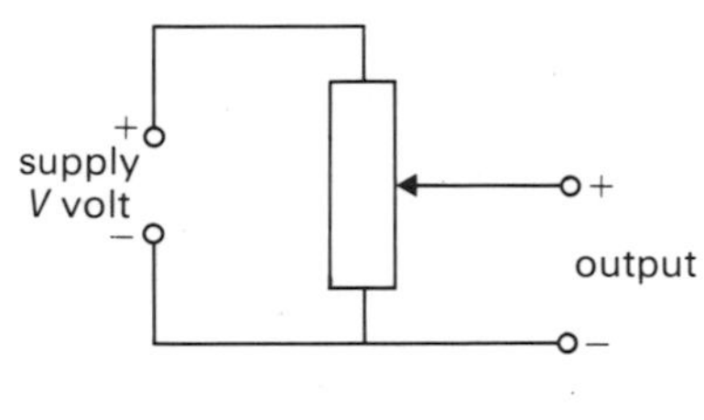

Figure 10.10 The potential divider

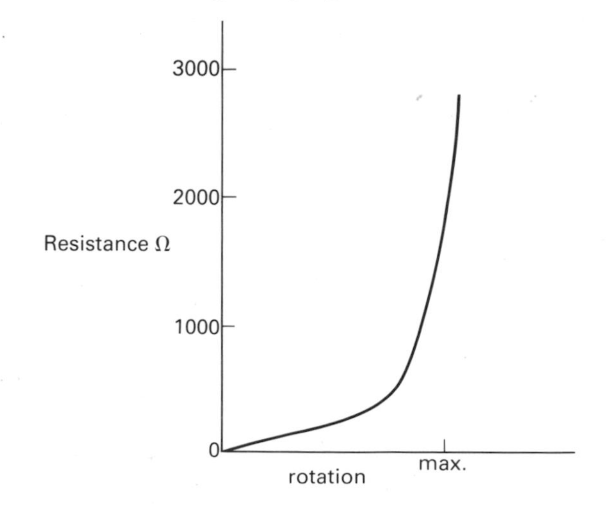

Figure 10.11 Resistance variation of a log pot

Using transistors for control of current

A transistor can form the basis of a very smooth method for controlling current; for a simple circuit, see unit thirteen (*figure 13.5*).

Unit 11 The amplifier

Amplifier gain

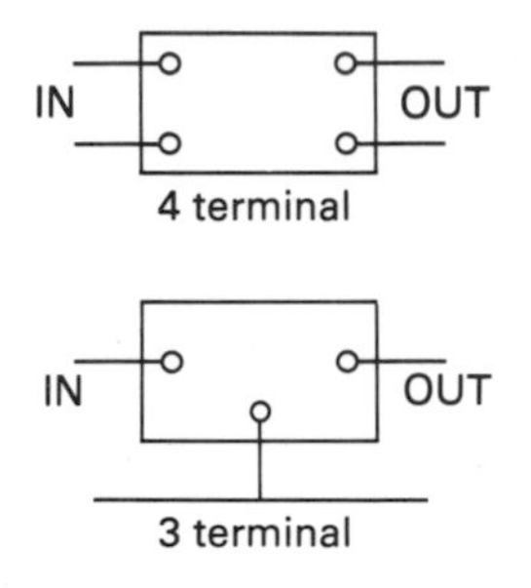

Figure 11.1
Black box amplifier symbols

For simplicity, at the moment, we will consider the amplifier to be a **black box**. This expression is used to describe any device whose action is going to be studied without going into the details of how it does it, or what it contains. The word **amplify** simply means **increase**, so we will now consider exactly what is being increased. The amplifier can be represented as a 4 terminal or 3 terminal black box, as in *figure 11.1*. For most purposes the three terminal version is satisfactory and will be used throughout.

For most people, experience with amplifiers has been as part of some electronic system in which a small voltage, as supplied by perhaps a microphone, is amplified and used to drive a loudspeaker. We describe this as **voltage amplification** and the increase is called the **voltage gain**:

$$\text{voltage gain} = \frac{\text{change in output voltage}}{\text{corresponding change in input voltage}}$$

Thus, if a microphone signal rises from 1 mV to 3 mV and the output voltage changes from 1·6 V to 2·4 V

$$\begin{aligned}\text{voltage gain} &= \frac{2{\cdot}4 - 1{\cdot}6}{0{\cdot}003 - 0{\cdot}001} \\ &= \frac{0{\cdot}8}{0{\cdot}002} \\ &= 400\end{aligned}$$

An amplifier will also amplify current, and the increase it produces is called the **current gain**:

$$\text{current gain} = \frac{\text{change in output current}}{\text{corresponding change in input current}}$$

The ordinary microphone amplifier is both a current and a voltage amplifier. Now because

$$\text{power in watts} = \text{voltage} \times \text{current in amps}$$

and since both the current and the voltage have increased, we must have more power coming out than was going in.

Therefore:

$$\text{power gain} = \text{voltage gain} \times \text{current gain}$$

No machine can give out more power than is put into it, so where does this additional power come from? It comes, of course, from the **power supply**, which every amplifier requires. The power supply may take the form of a battery, or be obtained via the mains. We will not show the power supply on our black box amplifiers since it is obvious that it must be present.

The signal source and load

Most of the signal sources we meet are alternating voltages, such as the output from microphone, pick-up or signal generator. The **signal generator** is a useful source because its voltage and frequency can be easily varied. All a.c. signal sources can be represented, as shown in *figure 11.2*, by an alternating e.m.f., V_s, in series with the source impedance, Z. Compare this with a d.c. source, which would be drawn as in *figure 11.3*.

The amplifier also has an input and output impedance and these may be shown inside the black box as in *figure 11.4*. When the amplifier is feeding its output into a **load**, the amplifier itself can be thought of as an a.c. signal source of voltage V and impedance Z_{out} as shown in *figure 11.5*. Combining all these ideas gives the complete black box amplifier shown in *figure 11.6*.

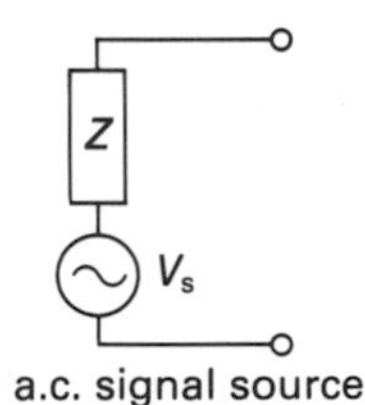

Figure 11.2 General method of representing an a.c. signal source

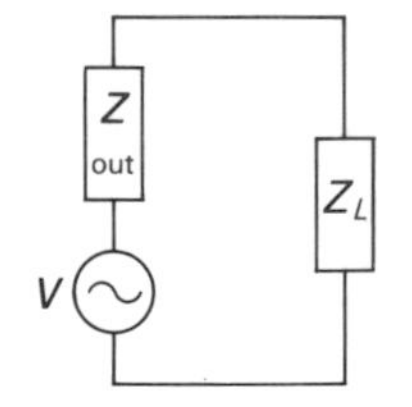

Figure 11.5
Connecting amplifier output to a load

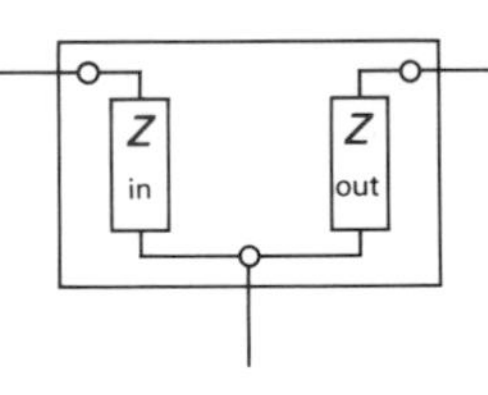

Figure 11.4
Input and output impedance shown inside the black box

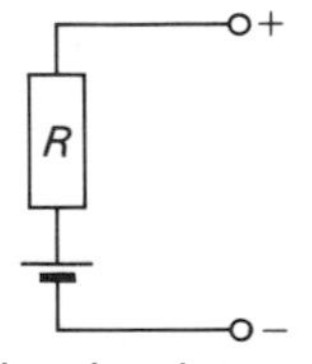

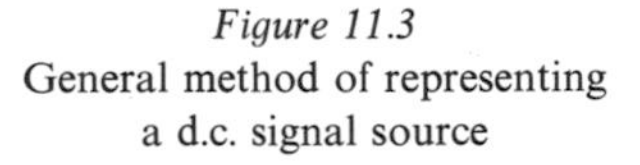

Figure 11.3
General method of representing a d.c. signal source

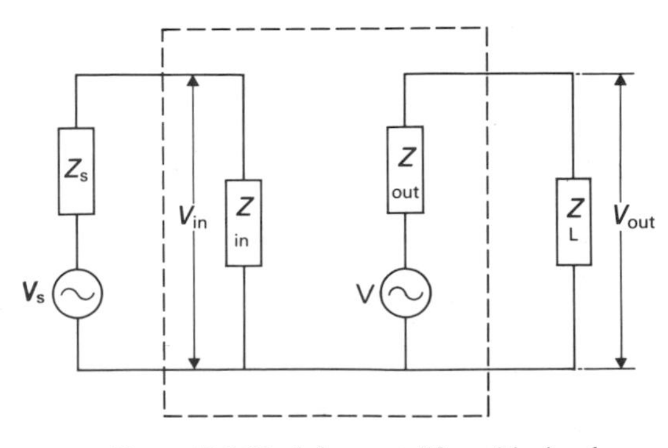

Figure 11.6 Black box amplifier with signal and load connected

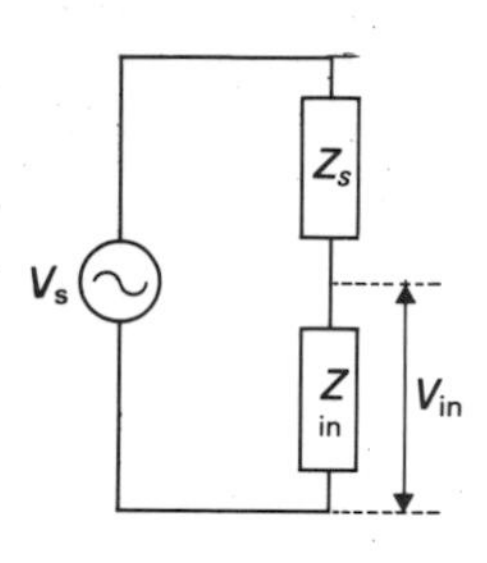

Figure 11.7
The input potential divider

Matching the signal source

At the input to an amplifier we would normally wish to transfer maximum voltage from the source to the input. This can be done by arranging for the source impedance to be much less than the amplifier input impedance. If the two impedances are redrawn, as in *figure 11.7*, it becomes more obvious that they form a potential divider across the source. We can then calculate the actual voltage being applied to the input terminals of the amplifier thus:

$$V_{in} = \frac{Z_{in}}{Z_s + Z_{in}} \times V_s$$

Example If $V_s = 300\,\text{mV}$, $Z_s = 3000\,\Omega$ and $Z_{in} = 1500\,\Omega$

$$V_{in} = \frac{1500}{4500} \times 300\,\text{mV}$$

$$= \frac{1}{3} \times 300\,\text{mV}$$

$$V_{in} = 100\,\text{mV}$$

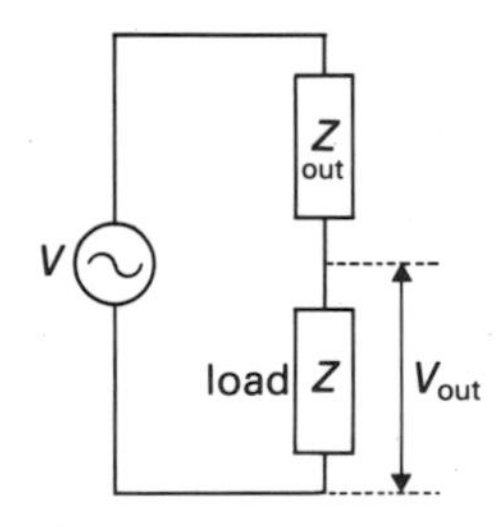

Figure 11.8
The output potential divider

Matching the load

The maximum power theorem tells us that the output impedance of the amplifier must equal the load impedance in order to obtain maximum transfer of power to the load. *Figure 11.8* shows how the output impedances act as a potential divider. The voltage at the output terminals can be calculated by using the equation:

$$V_{out} = \frac{Z_L}{Z_{out} + Z_L} \times V$$

It is much more important this time to make $Z_{out} = Z_L$ so that the amplifier delivers maximum power into the load, which is usually a loudspeaker.

If $Z_L = Z_{out} = 3\,\Omega$, and the voltage gain of the amplifier is 40, we can now complete the calculation begun above and calculate the power delivered to the $3\,\Omega$ loudspeaker.

$$V_{in} = 100\,\text{mV (calculated above)}$$

$$V = 40 \times 100\,\text{mV} \quad = 4\,\text{V}$$

$$V_{out} = \frac{3}{3+3} \times 4 \quad = 2\,\text{V}$$

$$\text{power} = \frac{V^2}{Z_L} = \frac{2 \times 2}{3} \qquad \text{power} = 1{\cdot}3\,\text{W}$$

Joining amplifiers together

The black box amplifier we have just described probably contains several stages of amplification. Each stage must be coupled to the next stage so that the a.c. signal is successively amplified by each stage. The three methods of coupling are described below.

Direct coupling

This method is very commonly used but is not easy to handle and is beyond the scope of this book. It must be used if d.c. or low frequency a.c. signals are to be amplified.

Capacitor coupling

This is the simplest method to follow, and the arrangement is shown in *figure 11.9*. The value of the coupling capacitor should be chosen so that it has a low reactance, to provide an easy path for the a.c. signal to the next stage. The effect of the coupling capacitor C_c and the input capacitance C_i of the next stage is shown on the graph (*figure 11.10*).

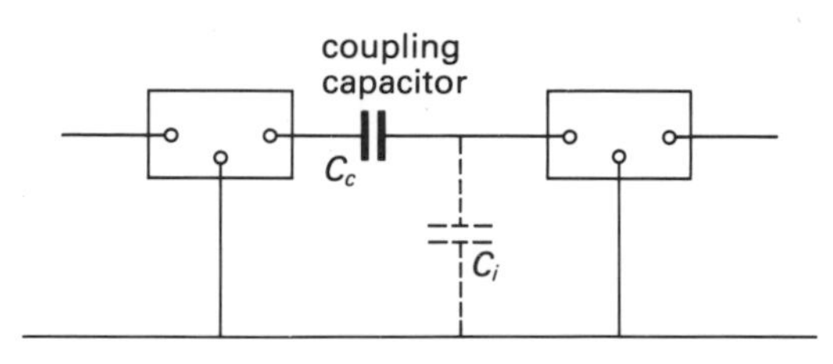

Figure 11.9
Capacitor coupling between two amplifiers

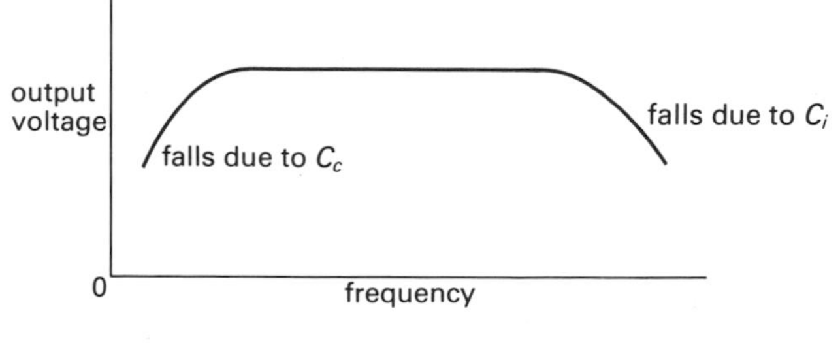

Figure 11.10
Variation of output with frequency

Even if a suitable value for the coupling capacitor is chosen, there may be another cause of signal loss; the output impedance of the first stage is usually greater than the input impedance of the next. There are two ways of dealing with this signal loss:

a use an extra stage of amplification to overcome the loss.

b alter the input impedance so that it matches the output impedance of the previous stage.

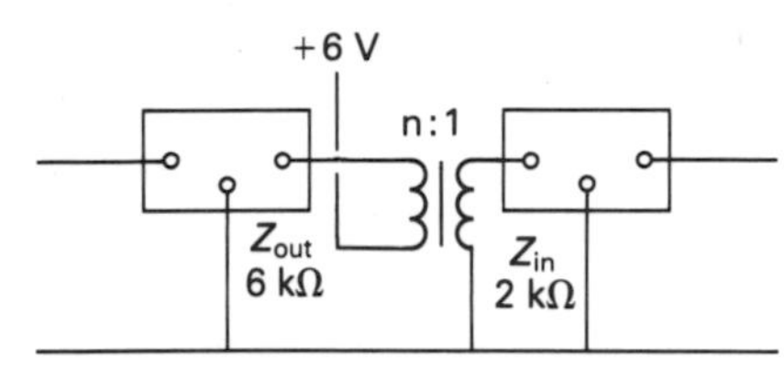

Figure 11.11
Transformer coupling between two amplifiers

Transformer coupling

Figure 11.11 shows how a transformer can be used to couple two amplifier stages.

If $Z_{out} = 6\,k\Omega$, $Z_{in} = 2\,k\Omega$ and turns ratio $= n{:}1$

then $$\frac{6\,k\Omega}{2\,k\Omega} = \left(\frac{n}{1}\right)^2 \quad \text{(see unit five)}$$

$$n^2 = 3 \qquad \therefore \quad n = 1{\cdot}73$$

This method of coupling is rarely used at audio frequencies, because the transformer would need quite a large iron core to do the job properly. At radio frequencies, transformer coupling is very frequently used, the radio frequency transformer being quite small and inexpensive.

Feedback

The word **feedback** is used to describe the process whereby all or part of the output generated by a system is fed back to the input of that system. It is a very general idea and certainly not confined to electronics. In human relations, for example, you may see from a person's face that the words which you are speaking are not having the desired effect, so you change your output accordingly. If the feedback is **positive**, it reinforces or strengthens what we are doing, resulting in continued or even greater output. If the feedback is **negative**, it will tend to cancel or reduce the effect which produced it. A simple electrical feedback system is represented by the room thermostat which turns the heat on or off according to the temperature.

Positive feedback

A common example of positive feedback is experienced when a microphone is feeding an amplified signal to a loudspeaker in the same room. If the microphone picks up too much of the output from the speaker a characteristic howl is set up.

If it is required to generate electrical oscillations, positive feedback is used to keep the circuit oscillating (unit seventeen).

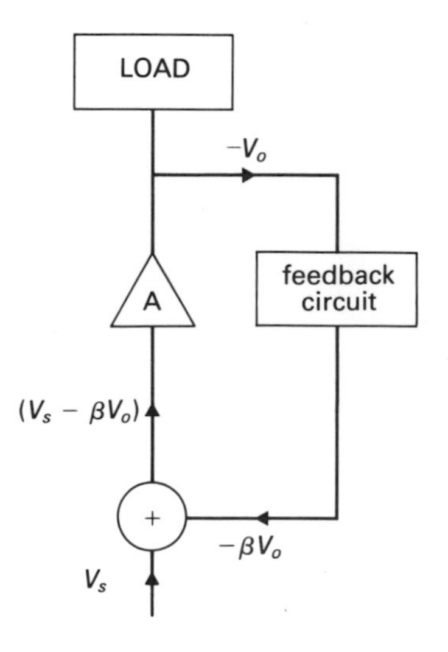

Figure 11.12 An amplifier with negative feedback

Negative feedback

The simplest way of drawing an amplifier with negative feedback is shown in *figure 11.12*. Without the feedback, the output voltage V_o is A times the input voltage V_s

$$V_o = A\,V_s$$

A is called the **open loop gain**. With feedback, the input voltage is reduced to $(V_s - \beta\,V_o)$ where β is a number less than 1 and fixed by the feedback circuit. This time:

$$V_o = A(V_s - \beta\,V_o)$$

It must seem a little pointless at this moment to have an amplifier and then discuss how to reduce its amplification! There are, however, three great advantages in applying negative feedback:

a the amplifier will amplify a wider range of frequencies by the same amount (see *figure 11.13*).

b distortion is reduced, so the output waves should have the same shape as the input waves.

c the gain of the amplifier changes very little as the components alter their performance with age.

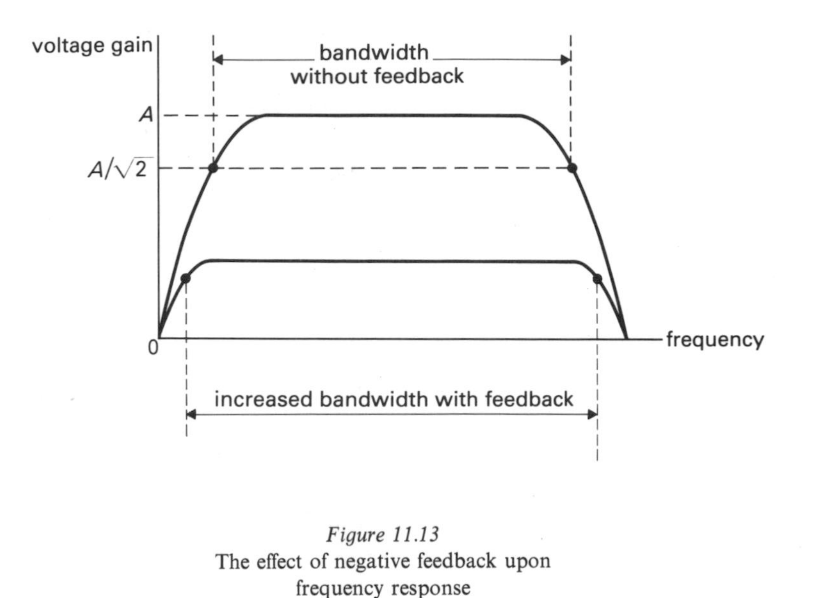

Figure 11.13
The effect of negative feedback upon frequency response

Frequency response of an amplifier

If the input voltage to the amplifier is kept constant and the output voltage is measured over a wide range of frequencies, the resulting graph would look like *figure 11.13*. The two points on the curve at which the gain has fallen to $A/\sqrt{2}$ are called the 3 dB points. The decibel (dB) scale compares one signal with another and a fall of 3 dB means the signal voltage has fallen to 0·707 (or $1/\sqrt{2}$) of its original value. Three decibels represents a drop in output which the ear would notice as being significant. The range of frequencies between the 3 dB points is called the **bandwidth** of the amplifier. The bandwidth is increased by applying negative feedback and the resulting loss in gain must be made up by using extra stages of amplification, so the open loop gain must be increased.

Class of amplification

The only type of amplification considered in this book is called class A. In order to see the significance of this label, here is a brief description of the three classes of amplification:

Class A The output current waveform has exactly the same shape as the input waveform. The output is said to be undistorted in this case.

Class B No output current flows during the negative half of the input cycle. The output is therefore half wave rectified and needs another class B amplifier to provide the other half if an undistorted wave is required.

Class C The output current flows in a brief pulse at the positive peaks of the input voltage.

The operational amplifier

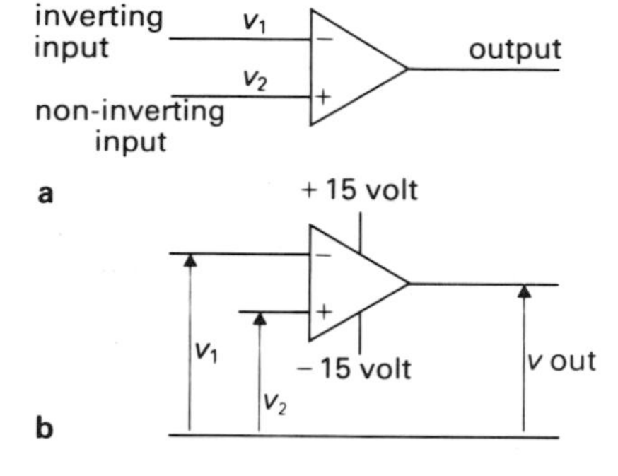

Figure 11.14a Operational amplifier symbol

Figure 11.14b Operational amplifier with connections to power supply

The operational amplifier is a high gain, direct coupled voltage amplifier; it has one output but two inputs, as shown in the circuit symbol in *figure 11.14*. This device will amplify the difference in the voltages applied to these two inputs, so the output will be as follows:

If $V_1 = V_2$ the output will be zero.
If V_2 is more positive than V_1 the output rises above zero.
If V_1 is more positive than V_2 the output falls below zero.

The greater the difference between the input signals, the greater the output voltage becomes until the amplifier is driven to saturation, as shown in *figure 11.15*.To summarise this we could write:

$V_{out} = (V_2 - V_1) \times \text{gain}.$

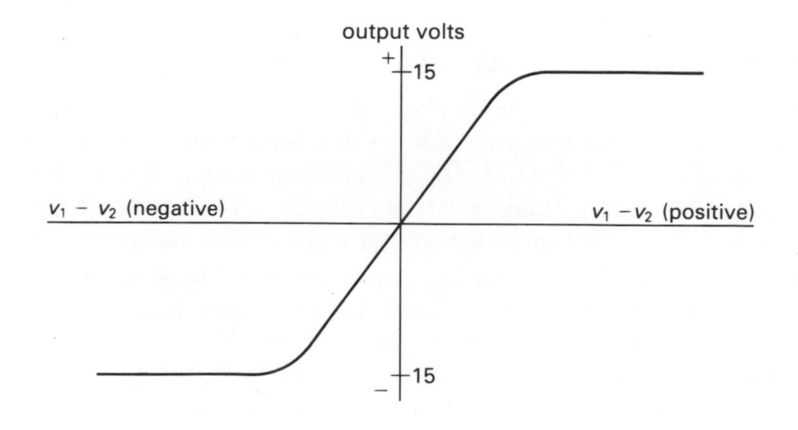

Figure 11.15
Variation of output voltage with input

In order to obtain a single input terminal, one of the input terminals is grounded (*figure 11.16*).

a is an **inverting amplifier**, so as the input voltage V_{in} rises, the output voltage becomes increasingly negative i.e. it falls further and further below zero.

b is a **non-inverting amplifier**, so as the input voltage rises, the output voltage rises increasingly above zero.

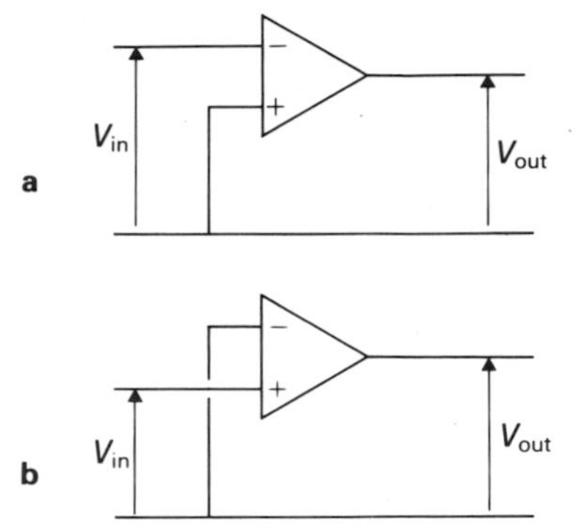

Figure 11.16a Using the inverting input only

Figure 11.16b Using the non-inverting input only

The simple operational amplifier has the following input and output characteristics:

a a very high voltage gain of around 100 000.
b an input impedance of about 1 MΩ.
c an output impedance of about 100 Ω.
d the range of frequencies which can be amplified is from zero (d.c.) to about 100kHz.
e they introduce rather a lot of noise if used to amplify signals of less than 100mV.

The gain of an operational amplifier can be adjusted by applying a controlled amount of negative feedback, by means of the resistor R_f (*figure 11.17*). Notice that the feedback resistor returns the signal to the **inverting** input, so providing **negative** feedback. The closed loop gain and input impedance are calculated as follows:

$$\text{closed loop gain} = \frac{V_{out}}{V_{in}} = -\frac{R_f}{R_i}$$

(the − sign is simply to indicate that the inverting input is in use)

input impedance = R_i

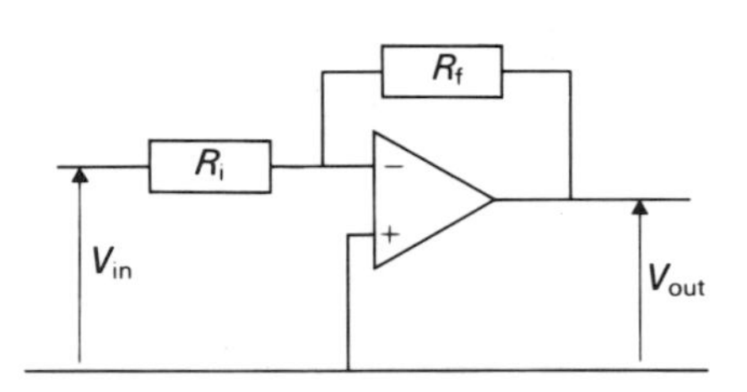

Figure 11.17 Using negative feedback to control gain

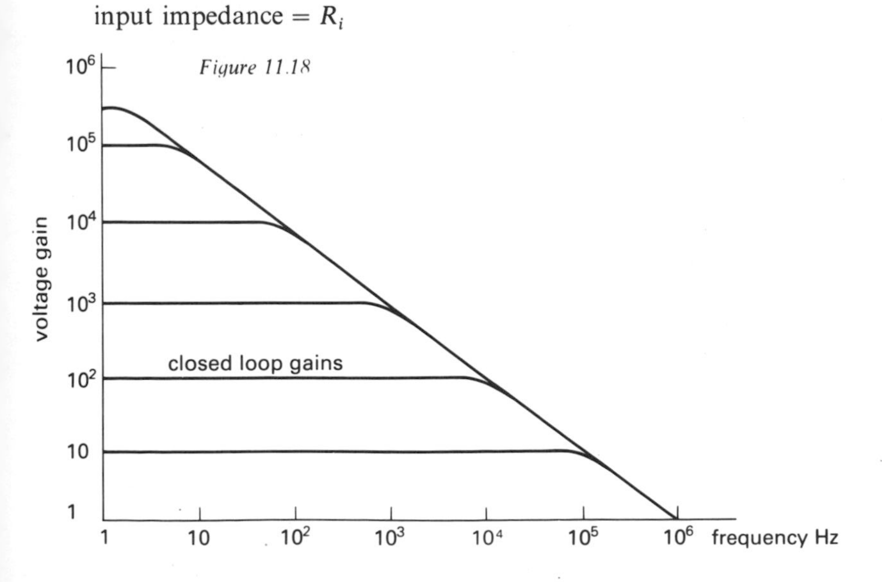

The greater the amount of negative feedback, the lower the closed loop gain and the greater the bandwidth (*figure 11.18*). Notice that the closed loop gain multiplied by the maximum frequency response is always equal to 1 million in this particular example.

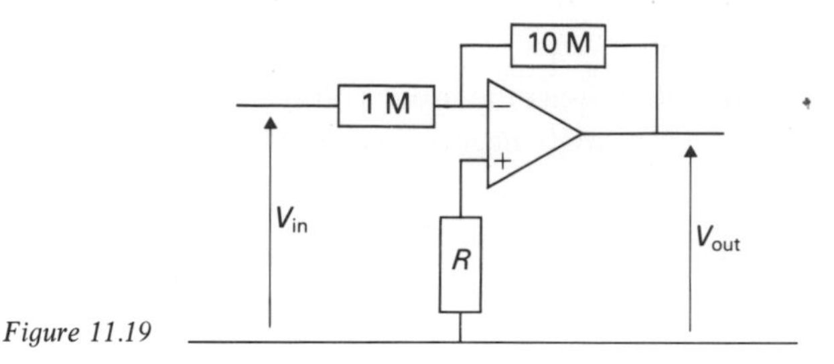

Figure 11.19

The circuit shown in *figure 11.19* shows how a voltage gain of 10 may be obtained. The resistor labelled R should have a value equivalent to the 1 MΩ and 10 MΩ resistors in parallel.

$$R = \frac{1 \times 10}{1 + 10} = 0{\cdot}909\ \text{M}\Omega$$

So the nearest preferred value would be 910 kΩ. The purpose of this resistor is to prevent a steady d.c. output voltage due to the difference in voltages at the two inputs, and also to reduce the effect of temperature changes.

Simple theory of the operational amplifier

For the ideal operational amplifier, the input impedance is infinite, hence no input current flows into the amplifier, and the entire input current i flows through the feed back resistor R_f, (*figure 11.20*).

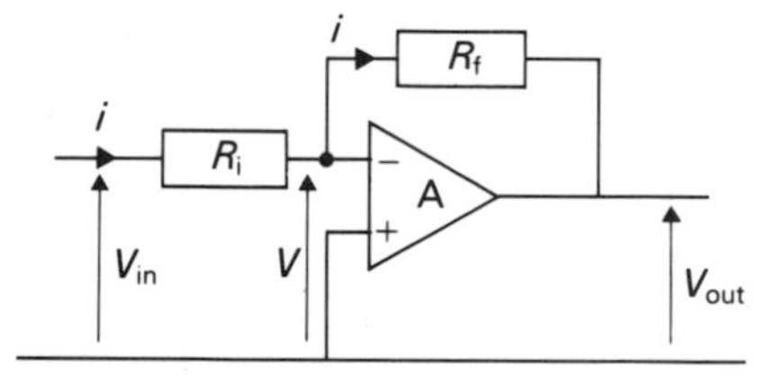

Figure 11.20
The path of the input current to an operational amplifier

Example If $V_{out} = 15$ V (it cannot be greater than the supply voltage)

and $A = 100\,000$ (for the ideal operational amplifier, A = infinity)

then $V = \dfrac{15}{100\,000} = 0{\cdot}00015$ V

Whenever we put A = infinity, we are using what is known as the **infinite gain approximation.**

This calculation shows that the voltage at the input terminal is virtually zero, hence this terminal is said to be a **virtual earth.** Now apply Ohm's law to R_i:

$$\text{p.d. across } R_i = (V_{in} - V) = V_{in}$$

$$\therefore \quad i = \frac{V_{in}}{R_i}$$

But if the input impedance for the whole amplifier is Z_{in} then:

$$i = \frac{V_{in}}{Z_{in}}$$

so this means that $Z_{in} = R_i$

Now apply Ohm's law to R_i and R_f remembering that the input terminal is virtually at zero volts:

p.d. across $R_i = V_{in}$ p.d. across $R_f = V_{out}$

$$V_{in} = i R_i \qquad V_{out} = -i R_f$$

$$\therefore \quad i = \frac{V_{in}}{R_i} \qquad \therefore \quad V_{out} = -\frac{V_{in}}{R_i} R_f$$

$$\therefore \quad V_{out} = -\frac{R_f}{R_i} \times V_{in}$$

As can be seen from this final expression, the closed loop gain is equal to:

$$-\frac{R_f}{R_i}$$

The integrator

If a capacitor is used as the feedback component in an operational amplifier, as in *figure 11.21*, then for a constant value of V_{in}, V_{out} will increase at a steady rate until the amplifier saturates (*figure 11.22*). The reason for this can be seen if we consider how the capacitor is being charged. Back in unit three, *figure 3.10b* shows the current falling as the charge on the capacitor increases. In the operational amplifier circuit, the charging current remains constant at a value *i*, so the charge increases at a uniform rate. It should be noted that the charge on *C* builds up much more slowly in this circuit than if V_{in} was simply charging *C* via *R*. The time constant, *CR*, is in fact increased to *ACR*, where *A* is the gain of the amplifier.

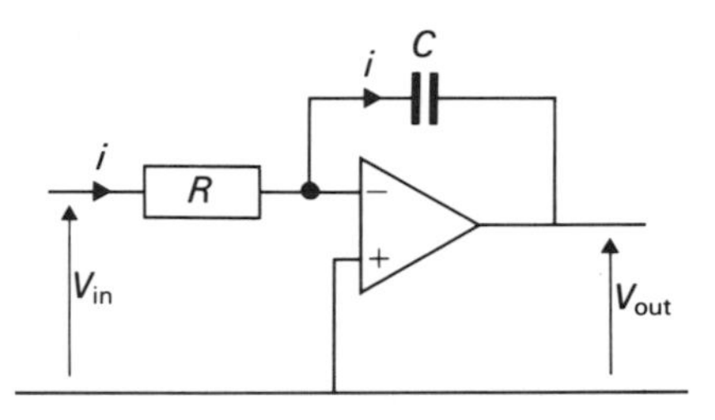

Figure 11.21 The integrator

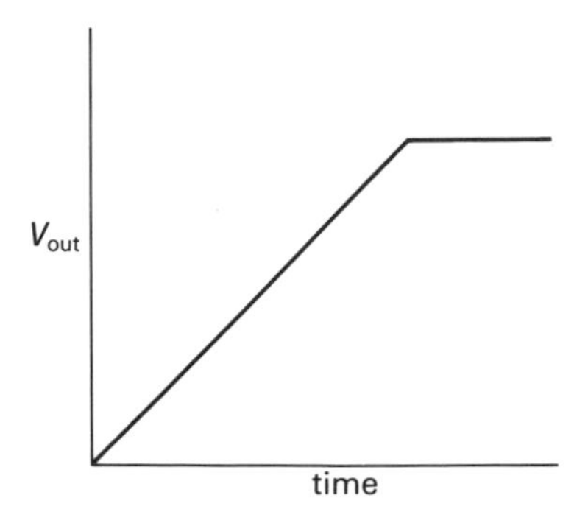

Figure 11.22
Ramp voltage produced by the integrator

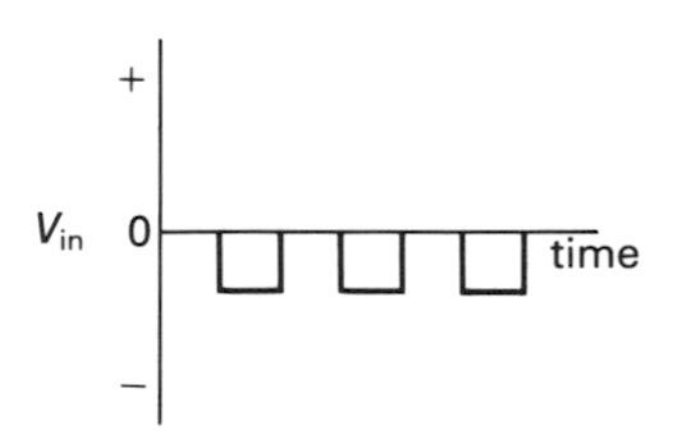

Figure 11.23
Square wave input to integrator

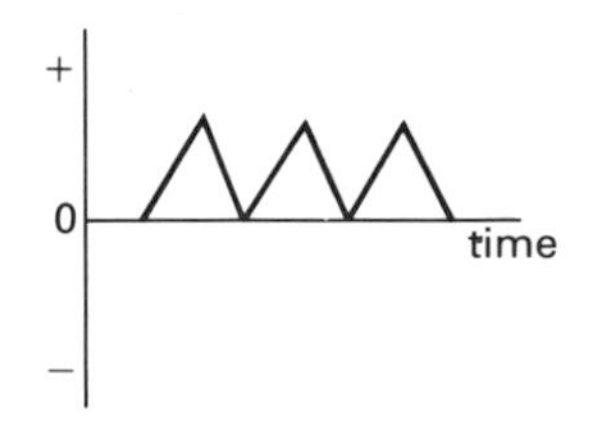

Figure 11.24
Resulting output from integrator

If the input is switched alternately high and low by applying a square wave, the output ramp will alternately rise and fall, as shown in *figure 11.23–24*.

Adding with an operational amplifier

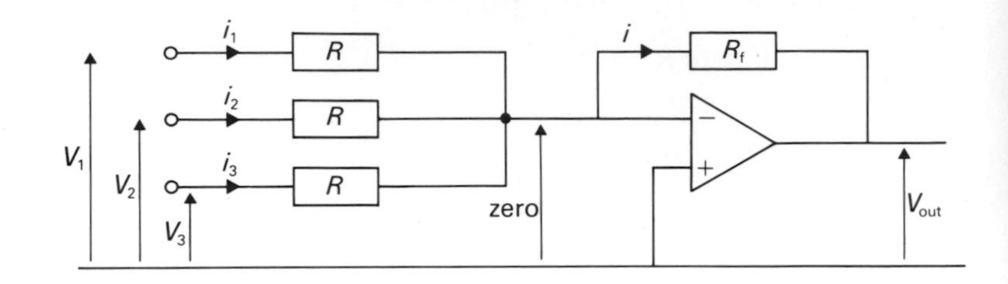

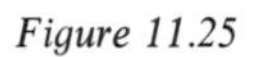
Figure 11.25

If each of the three inputs shown in *figure 11.25* produces a current i_1, i_2 and i_3, then the total current, i, will be equal to their sum:

$$i = i_1 + i_2 + i_3$$

But by applying Ohm's law to the input resistances:

$$i_1 = \frac{V_1}{R} \qquad i_2 = \frac{V_2}{R} \qquad i_3 = \frac{V_3}{R}$$

and applying Ohm's law to the feedback resistor:

$$i = -\frac{V_{out}}{R_f}$$

$$\therefore \quad \frac{V_1}{R} + \frac{V_2}{R} + \frac{V_3}{R} = -\frac{V_{out}}{R_f}$$

This means that the output voltage will be proportional to the sum of the input voltages.

Semiconductors

Unit 12

Solid metals are crystalline substances, and each tiny crystal consists of atoms arranged in a regular pattern called a crystal lattice (*figure 12.1*). If the metal is a good conductor of electricity, it is because the electrons furthest from the nucleus are easily detached at room temperature. These electrons can drift freely through the lattice when a p.d. is applied to the metal. Copper and silver are among the best conductors of electricity.

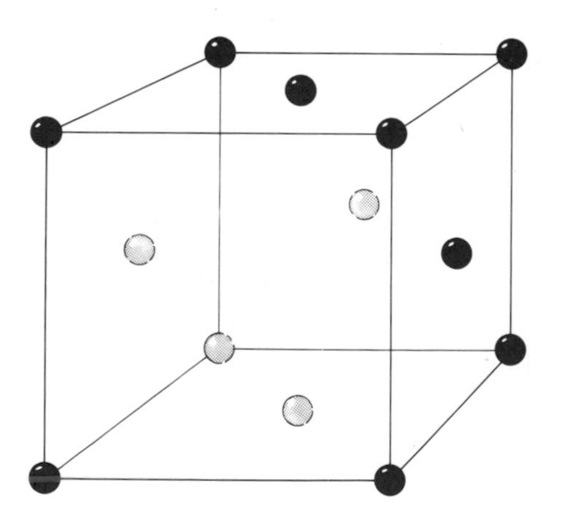

Figure 12.1
Arrangement of atoms in a crystal lattice

Insulators are made from materials in which the electrons are tightly bonded to the nucleus and a very high p.d. would be necessary to detach them. Many plastics, such as polythene and PVC are excellent insulators.

Between these two extremes there is a class of materials known as **semiconductors**. Examples of such materials are:

copper oxide	selenium	gallium arsenide
cadmium sulphide	germanium	silicon

lead sulphide (the material of the old 'cat's whisker' crystal set)

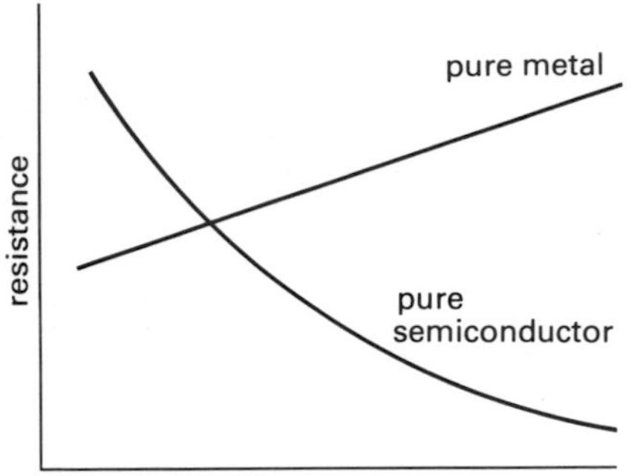

Figure 12.2 Graphs to show variation of resistance with temperature

The effect of temperature upon conduction in metals

As the temperature of a pure metallic conductor rises, there is an increase in the vibration of the atoms in the crystal lattice. This results in a greater chance of the free electrons colliding with an atom and being scattered out of their original path. So as the temperature rises, the metal becomes a slightly less effective conductor. The resistance of a pure metal thus rises with an increase in temperature. It also follows that the resistance falls as the temperature falls, and at very low temperatures the resistance may completely vanish. Such a metal is then said to be **superconducting**.

Intrinsic conduction

The lowest temperature which can exist is called absolute zero, and its value is about $-273°C$. A semiconductor at absolute zero has no free electrons, so cannot conduct; it is a good insulator. As the temperature of a semiconductor rises, the atoms begin to vibrate and shake out some free electrons, called **thermal electrons**, which can carry a charge through the semiconductor. Since the number of thermal electrons will increase as the temperature rises, it will become a better conductor, so its resistance will fall. A pure semiconductor, which conducts **only** by means of thermal electrons, is called an **intrinsic semiconductor.**

Electrons and holes

Figure 12.3 illustrates the formation of a free electron and hole in an intrinsic semiconductor. Both the electron and the hole are able to carry charges. When an electric field is applied, the electron carries a negative charge in one direction, while the hole carries a positive charge in the opposite direction. The holes are just the vacant places created by missing electrons. A hole can be filled by a lattice electron which has in turn left a hole (*figure 12.4*). This gives the impression of the hole having moved.

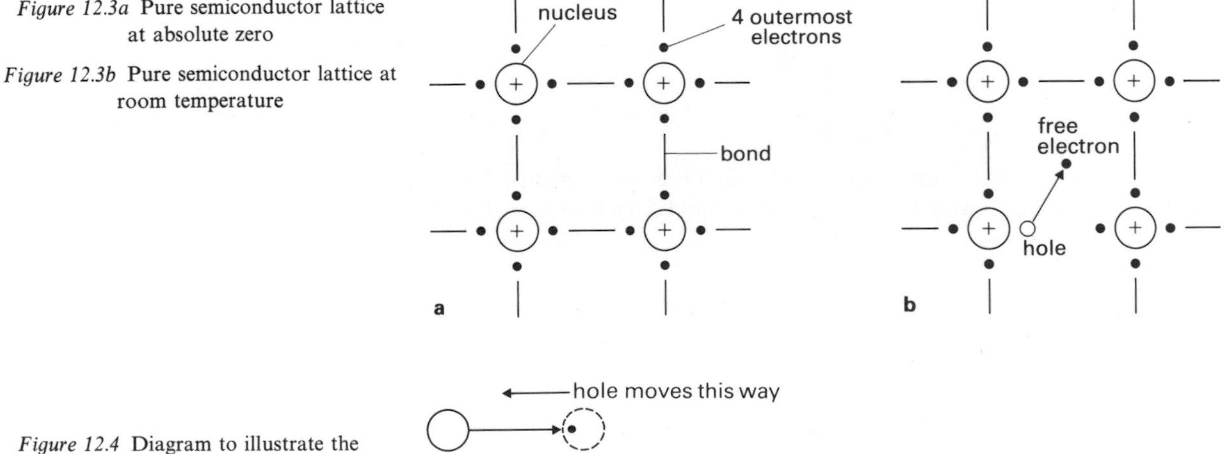

Figure 12.3a Pure semiconductor lattice at absolute zero

Figure 12.3b Pure semiconductor lattice at room temperature

Figure 12.4 Diagram to illustrate the apparent movement of a hole

Impurity conduction

An element in group five of the periodic table, such as phosphorus, arsenic or antimony, has five valence (bonding) electrons on the outside of the atom, furthest from the nucleus. If such an element is added in **very small** quantities to pure silicon (or germanium), it can enter the crystal lattice in place of a silicon atom, with no resulting distortion of the lattice pattern. The result of this is a large increase in the number of free electrons, since each five-valent atom gives one free electron (*figure 12.5*). Although there may only be one impurity atom in one hundred million silicon atoms, this still represents a very large number of free electrons. Such a semiconductor is called ***n*-type**, because the majority of the current is carried by negative charges (electrons). The electrons in this case are called the **majority carriers** and the relatively few holes which have been left by thermal electrons are the **minority carriers**. The group five impurity atoms are called the **donor atoms**, because they donate electrons.

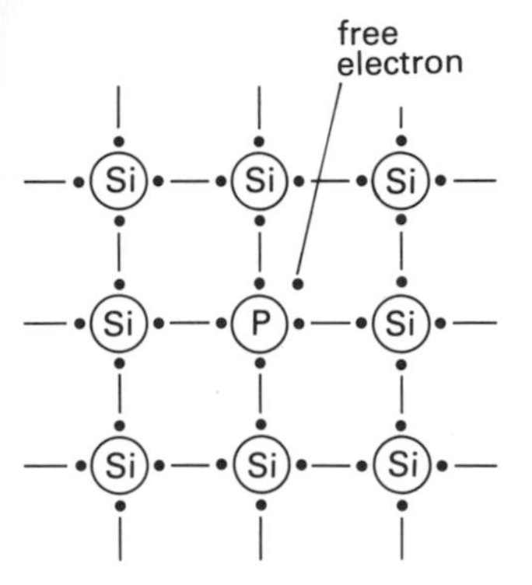

Figure 12.5 *n*-type silicon doped with phosphorus

An element in group three of the periodic table such as indium, boron or gallium has only three bonding electrons on the outside of the atom. If such an element is added to silicon in very tiny quantities, it enters the lattice, as before. This time, however, the result is one unpaired electron or hole for each impurity atom (*figure 12.6*). This type of silicon is called ***p*-type**, since the majority carrier is a positive charge, or hole. The group three impurity atom is called an **acceptor** atom because it accepts a valency electron from a neighbouring silicon atom. The minority carriers will be thermal electrons in this case.

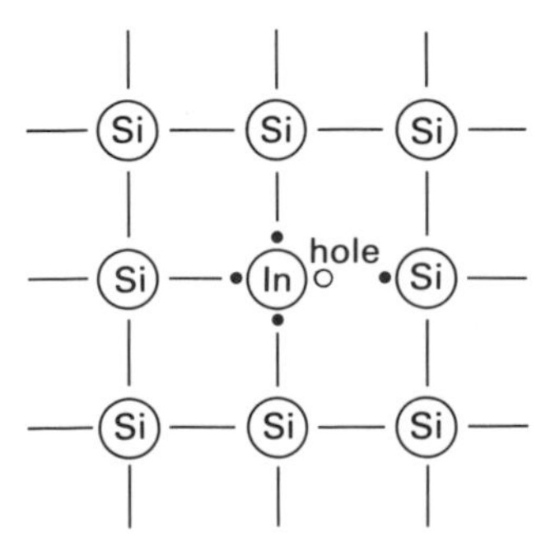

Figure 12.6 *p*-type silicon doped with indium

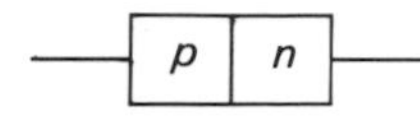

Figure 12.7 Schematic representation of a *p–n* junction

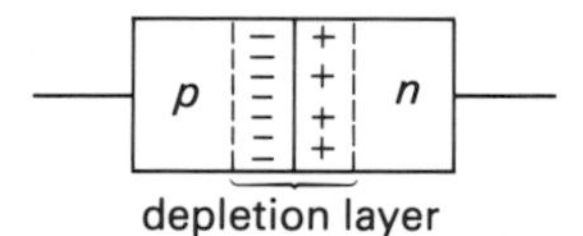

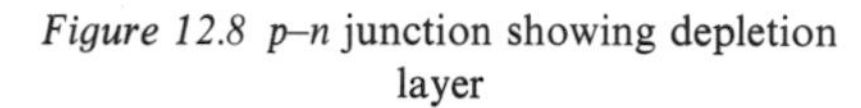

Figure 12.8 *p–n* junction showing depletion layer

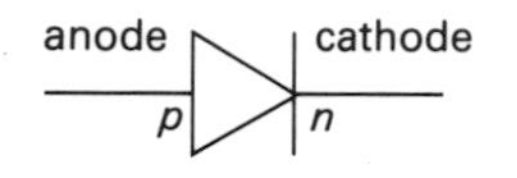

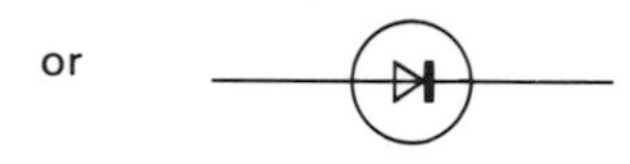

Figure 12.9 Circuit symbols for a *p–n* junction

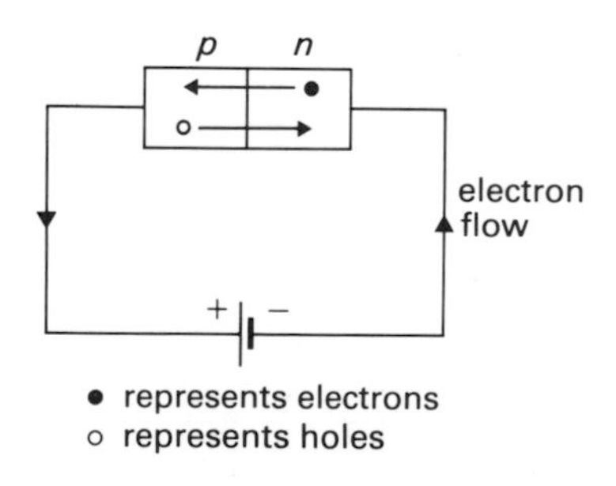

Figure 12.11 A *p–n* junction on forward bias

The *p-n* junction diode

The junction between *p*-type and *n*-type silicon or germanium has rather special properties. The junction must, however, start as a single crystal so that the crystal lattice is continuous. The *p* and *n* regions are then formed as shown diagrammatically in *figure 12.7*. As soon as the junction is created, electrons from the *n*-type spread into the *p*-type to fill its holes, and holes from the *p*-type spread into the *n*-type to be filled by electrons. This exchange does not go on for very long because a p.d. is set up by the exchange, so bringing it to a halt. *Figure 12.8* shows the **depletion layer** which is formed by this brief exchange of electrons and holes. Its thickness is only about 0.01 mm, and the p.d. across it is about 0.1 V for germanium and 0.6 V for silicon. The junction is called a *p–n* junction. One *p–n* junction will make a *p–n* junction diode, the symbol for which is shown in *figurè 12.9*. The outline shapes of some junction diodes are shown in *figure 12.10*.

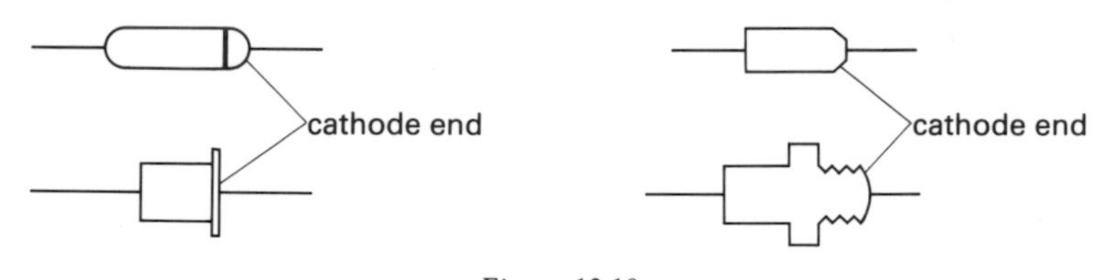

Figure 12.10
Outline shapes of some junction diodes

The *p-n* junction on forward bias

If a battery of e.m.f. greater than 0.6 V is connected across a *p–n* junction, as shown in *figure 12.11*, the *p*-type will be at a higher potential than the *n*-type, and holes and electrons can once again pass across the junction. The junction is said to be on **forward bias**, and its resistance is very low. The effect of forward bias on silicon and germanium diodes is shown in *figure 12.12*.

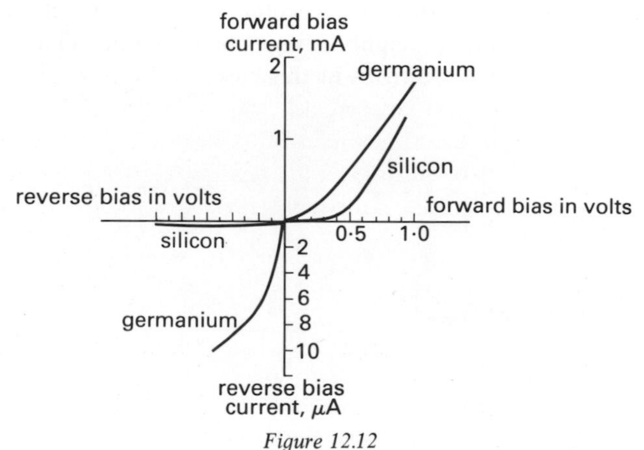

Figure 12.12
Comparison of germanium and silicon diodes

The *p*-*n* junction on reverse bias

A battery connected as shown in *figure 12.13* puts the diode on **reverse bias**; this simply has the effect of making the depletion layer wider. A very tiny current called the **leakage current**, will flow round the circuit. The leakage current is due to thermal electrons and holes, and will increase if the temperature rises. As the battery can only drive a tiny current, the resistance of a diode on reverse bias is very high.

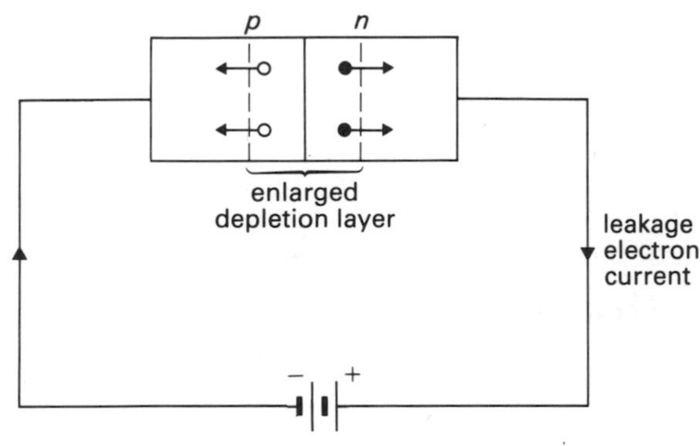

Figure 12.13 *p–n* junction on reverse bias

The capacitance of a *p*-*n* junction

Since the depletion layer is a good insulator, it behaves like the insulating material between the plates of a capacitor, with the *p*-type material and the *n*-type material acting as the two plates. A *p–n* junction diode does, therefore, have some capacitance. Furthermore, since the thickness of the depletion layer varies with the p.d. across the diode, its capacitance will vary according to the reverse bias voltage. Diodes specially made to be used in this way are called **varicap diodes**, and are used as the basis of the tuning circuit for changing channels on a television receiver.

The use of a diode as a rectifier

An alternating current (a.c.) is one which flows alternately in opposite directions. A direct current (d.c.) always flows in one direction. The process by which a.c. is changed into d.c. is called **rectification**, which may be either **half wave** or **full wave** rectification. A circuit which may be used for half wave rectification is shown in *figure 12.14*, and its effect is shown in *figure 12.15*. As the diode is alternately on forward and reverse bias, it will pass the top half of the a.c. wave and stop the bottom half.

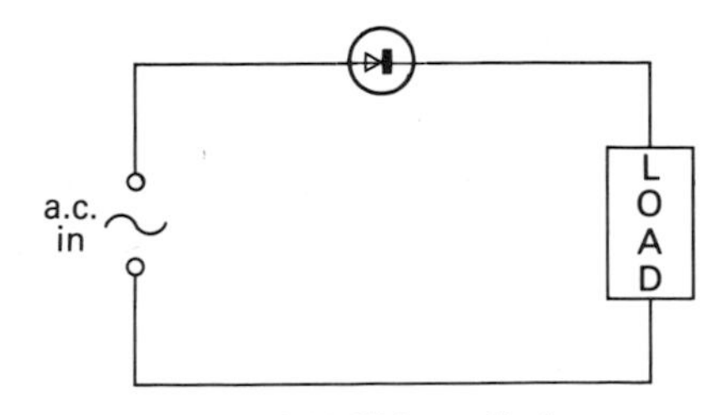

Figure 12.14 Using a diode as a half wave rectifier

Fig 1

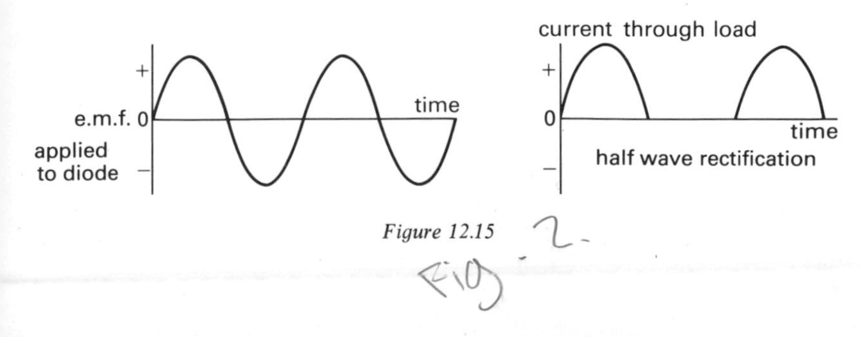

Figure 12.15 2.

Fig

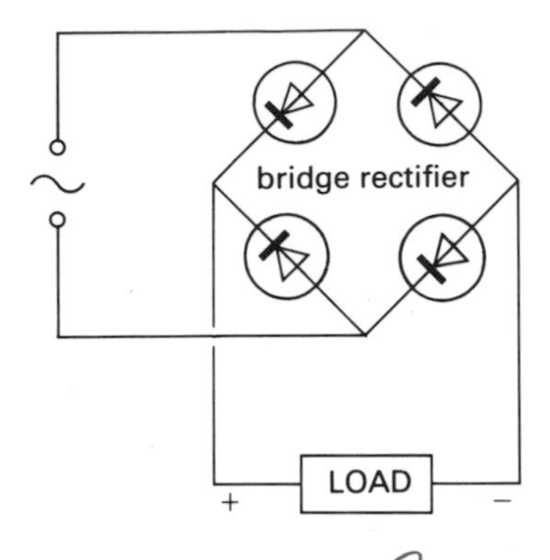

Figure 12.16
The arrangement of diodes in a full wave bridge rectifier

Although the result is d.c., it is better to call it **half wave rectified a.c.** *Figure 12.16* shows one method of producing **full wave rectified a.c.**, and *figure 12.17* shows how it differs from half wave rectification. It is easy to turn full wave rectified a.c. into smooth d.c. using the circuit shown in unit three (*figure 3.8* and *figure 3.16*).

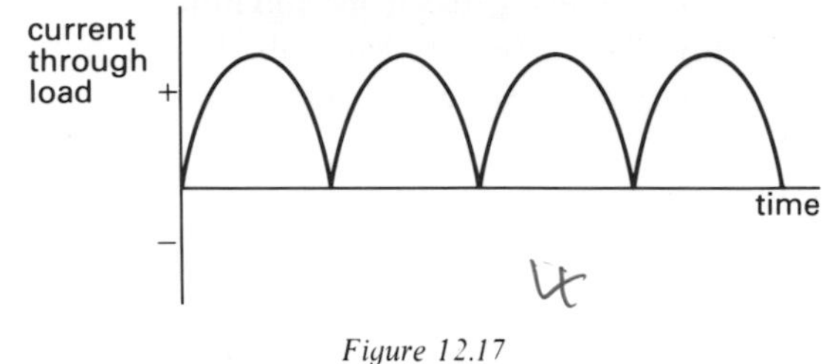

Figure 12.17
Output from a full wave rectifier

Zener diode

If the reverse bias on a *p-n* junction is increased sufficiently, the depletion layer eventually breaks down and a large current begins to flow. In the case of ordinary diodes, this breakdown is destructive and permanent, and the diode is made quite useless. If, however, the amount of impurity in the semiconductor is correct, the breakdown need not be permanent. In this case, as the reverse bias increases, the diode suddenly begins to conduct at what is called the **Zener point** (*figure 12.18*). The Zener point can be adjusted by the manufacturer to occur somewhere in the range 2·7 V to 72 V. It is also possible to buy a programmable Zener diode, in which the Zener point can be adjusted by the user, by means of a voltage applied to an extra lead on the diode.

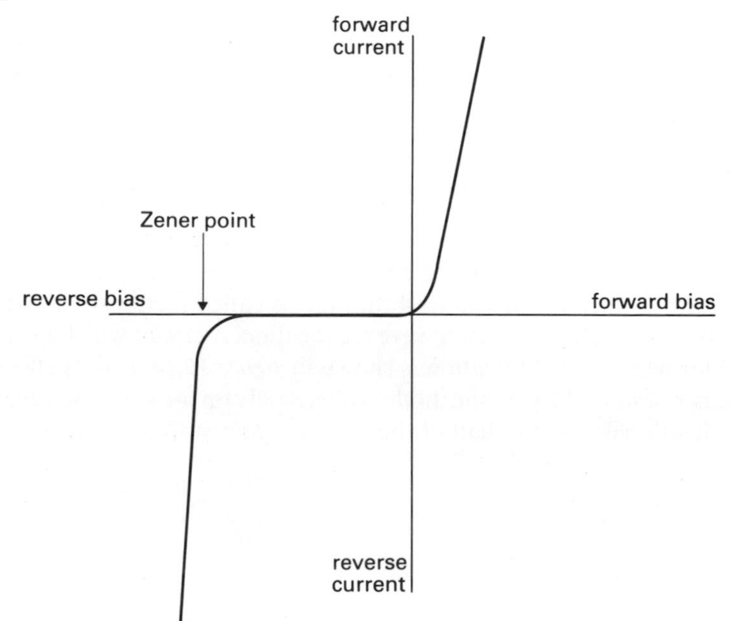

Figure 12.18
The characteristics of a Zener diode

When a Zener diode reaches its Zener point, the p.d. across it is called its **reference voltage**. If you have a Zener diode with 4 V7 written on it, its reference voltage is 4.7 V. This reference voltage will remain constant, even if the current through the diode increases. Care must be taken that the diode is not overheated by the current flowing through it, otherwise it will be destroyed. A small Zener diode may be rated at 400 mW and a larger one at perhaps 5 W.

A simple use for a Zener diode would be to supply a constant p.d. of, say, 5 V from a dry battery, whose e.m.f. falls continuously with use. A circuit is shown in *figure 12.19*. To work out a suitable value for R, the maximum current allowed through the Zener diode must first be calculated.

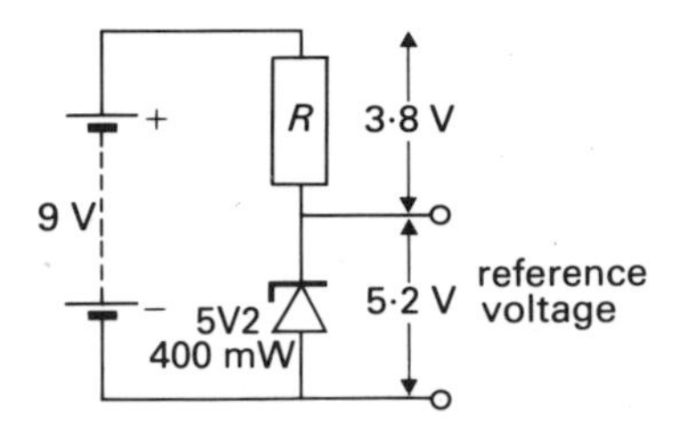

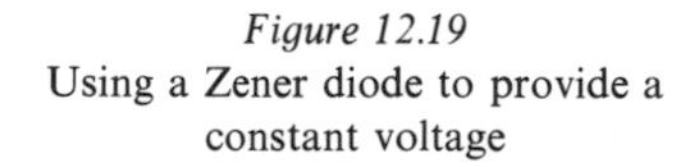

Figure 12.19
Using a Zener diode to provide a constant voltage

$$\text{power} = VI$$

$$0{\cdot}4 = 5{\cdot}2I$$

$$I = \frac{0{\cdot}4}{5{\cdot}2} = 0{\cdot}08\text{ A approximately}$$

then using $R = \dfrac{V}{I}$

$$R = \frac{3{\cdot}8}{0{\cdot}08}$$

$$R = 47\,\Omega$$

Photodiodes

When on reverse bias, the leakage current of a *p-n* diode is found to increase in proportion to the intensity of light falling on the junction. This effect can be used to make photometers (for measuring light intensity) and devices for 'reading' holes in punched cards and tapes.

As most semiconducting devices are influenced by light, it is usual to seal them in light-proof packages. The action of light upon a reversed biased diode can be seen by scraping the black paint from a glass encapsulated diode, such as an OA81 or OA202. An ohm-meter reading of the diode's resistance will be insensitive to light on forward bias, but sensitive to light when the ohm-meter connections are reversed.

Unit 13

The transistor

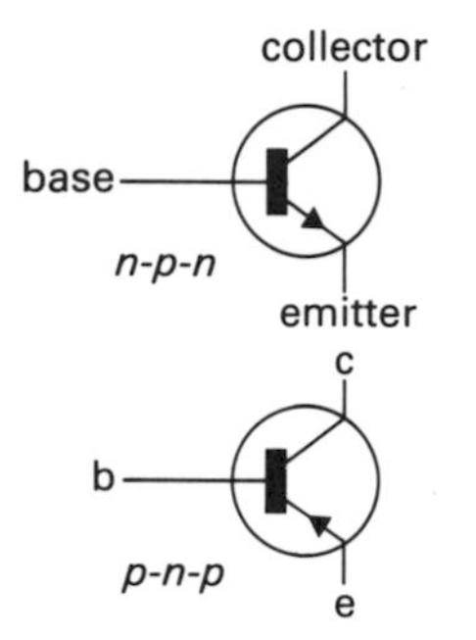

Figure 13.1 Transistor symbols

In 1948 it was discovered, in America, that a special arrangement of *p*-type and *n*-type semiconducting material was able to amplify a current. In 1950 the first **junction transistor** was produced by William Shockley.

In its simplest form, the transistor consists of two *p*-*n* junctions arranged as a sandwich, either *p*-*n*-*p*, or *n*-*p*-*n*. The semiconducting material may be germanium or silicon. Circuit symbols are shown in *figure 13.1*. The direction of the arrow on the emitter shows the direction of flow of conventional current (positive charge).

If you were to cut the top off a BC107 transistor case and look at the transistor inside, you would need a hand lens or a microscope to see very much. It is very obvious now that the transistor is a minute device and what we normally see is just the packaging, which makes it big enough to handle. The layers of *p*-type and *n*-type silicon in a transistor are shown, slightly simplified, in *figure 13.2*. There are essentially three uses for transistors:

a for controlling the strength of a direct current.
b acting as a fast switch.
c as an amplifier.

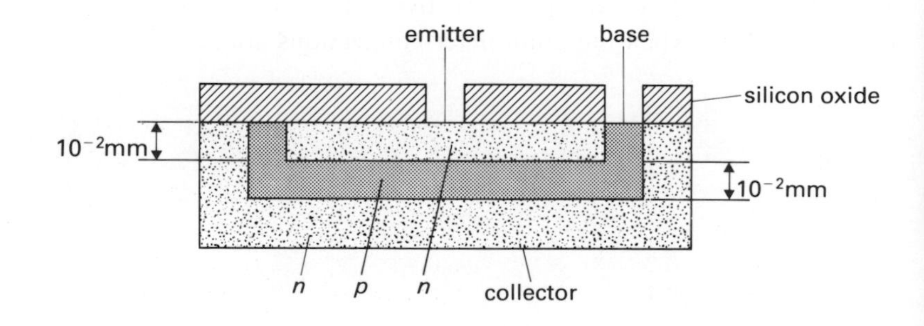

Figure 13.2
Arrangement of *p*- and *n*-type material in a transistor

Current control with a transistor

If the circuit shown in *figure 13.3* is set up, the base current to the transistor can be slowly increased. A graph of collector current plotted against base current is shown in *figure 13.4*. The results for two different transistors are shown on the same axes. These graphs clearly show that a small change in base current will produce a much larger change in collector current. The transistor, therefore, acts as a **current amplifier**.

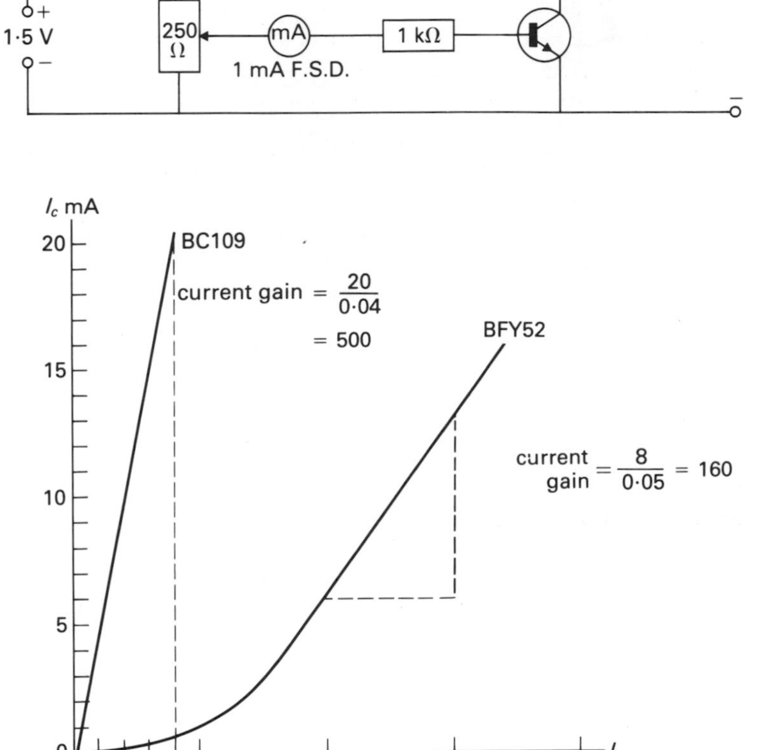

Figure 13.3 (*left*) Circuit used to see how collector current is changed by base current

Figure 13.4 (*left*) Transfer characteristics for two different transistors

Figure 13.5 shows how a transistor can be used to control current through a 6 V 40 mA lamp. If control of larger currents is required, a **power transistor** can be used. The 50 kΩ rheostat controls the very tiny base current. As the base current is increased, so the much larger collector current increases, and the lamp becomes brighter. The advantage of using a transistor in this way is that large currents can be smoothly controlled by a small, lower power rheostat. As the base current is so small, the rheostat does not become hot, but the transistor will become hot and care must be taken that it is not damaged. This can be done by fitting the transistor with a **heat sink**, which usually takes the form of blackened aluminium fins, to which the transistor is firmly fixed.

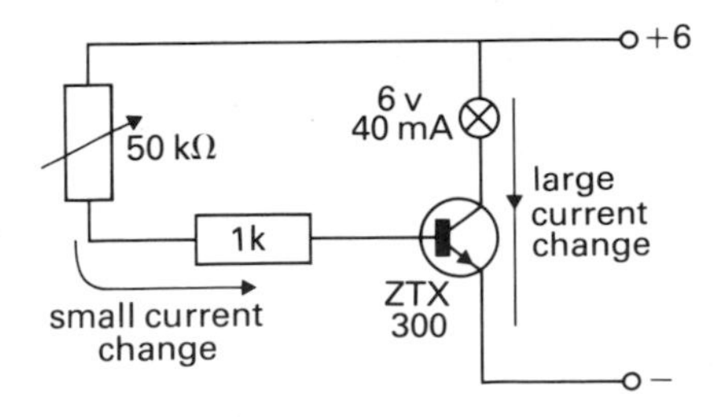

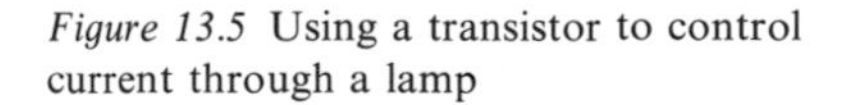
Figure 13.5 Using a transistor to control current through a lamp

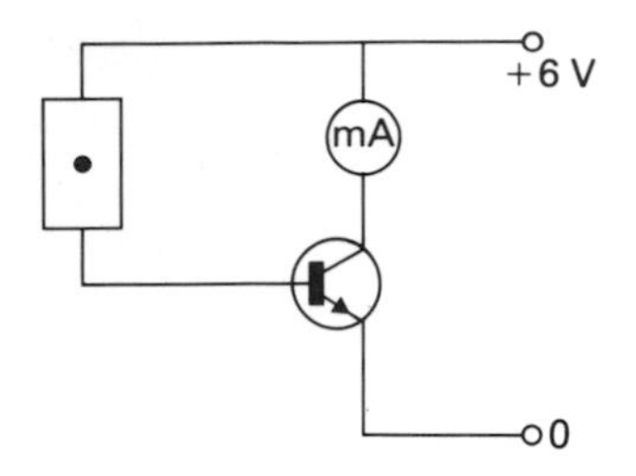

Figure 13.6 Using a thermistor to control base current

An interesting variation of the current control circuit is shown in *figure 13.6*, in which the base current is controlled by a thermistor. If the temperature of the thermistor rises, its resistance falls and the base current rises. This causes the collector current to rise and gives a higher reading on the meter. The meter could be scaled to read temperature, but the scale would not be uniform.

Transistor with cooling fins

The transistor as a switch

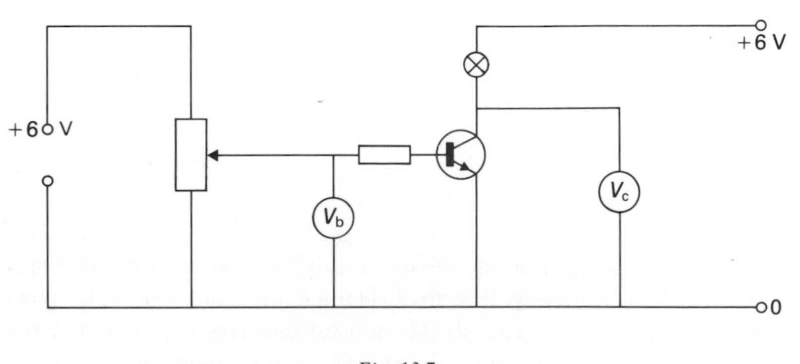

Fig. 13.7
To measure the effect of base voltage upon collector voltage

The circuit shown in *figure 13.7* enables the voltage present on the base to be altered. The effect of slowly raising the base voltage from zero is shown in *figure 13.8*. The lamp is off at the start because both ends of it are at +6 V, hence there is no p.d. to drive current through it. The

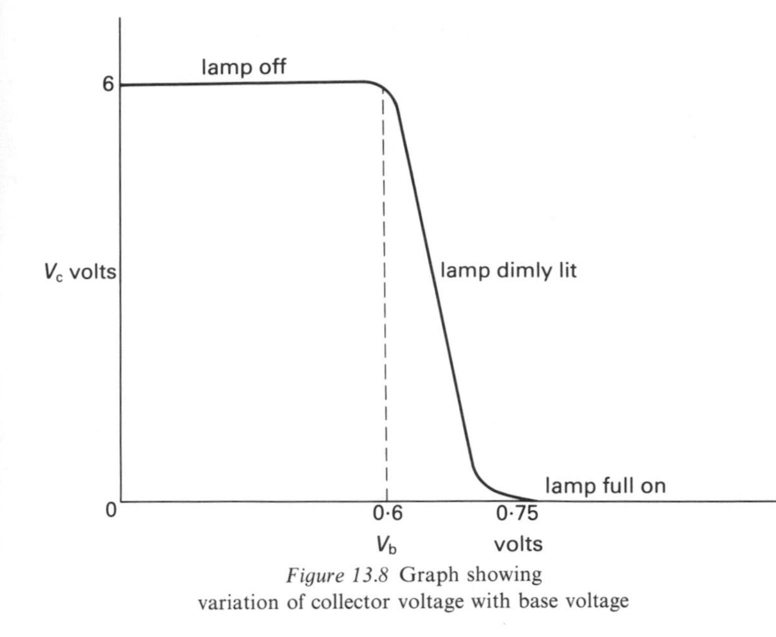

Figure 13.8 Graph showing variation of collector voltage with base voltage

transistor is said to be **off**, because it is not conducting along the collector–emitter path; it is behaving like an open switch (or, more accurately, like a high value resistor). When the base voltage reaches about 0.6 V, the transistor begins to conduct and current now flows along the collector-emitter path (look back to *figure 12.12* to see why). When the base voltage has risen to about 0.75 V, the transistor is said to be turned **on**; it is now behaving like a closed switch (or, rather, like a very low value resistor). As there is almost no resistance between the collector and the emitter, the collector will be at nearly the same voltage as the emitter, that is zero volts. The lamp now has a p.d. of 6 V across it, and lights fully. Notice that the base voltage has only to change from 0.6 V to 0.75 V to change the transistor from off to on.

Figure 13.9a shows an interesting way of changing the base voltage. The light dependent resistor and the 5K6 resistor act as a potential divider. In darkness the LDR resistance is high, so the base voltage is high and the transistor turned on. In bright light its resistance is low, so the base voltage is reduced and transistor is turned off. The lamp will light when the light dependent resistor is in darkness. Possible uses for such a circuit include automatic parking lights for cars, or automatic porch lights for the home.

One problem with the simple circuit in *figure 13.9a* is the possibility of the lamp being dimly lit, instead of fully on or fully off. This may not be much of a problem for a lamp but it would be much more serious if the transistor was operating a relay, which would tend to 'chatter' instead of switching smartly on and off. This problem can be overcome by using a two-transistor circuit known as a **Schmitt trigger**, as shown in *figure 13.9b*. The collector-emitter current for the second transistor, Tr_2, passes through the same resistor as the collector-emitter current for first transistor, Tr_1. Any change in Tr_2 current is therefore communicated back to Tr_1, and this tends to accelerate the change.

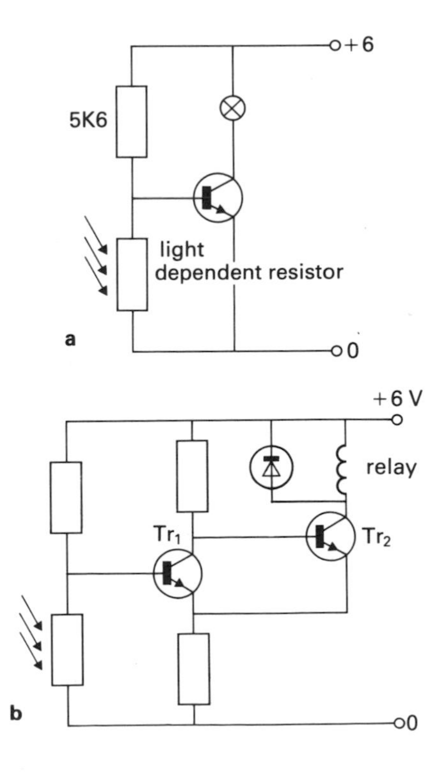

Figure 13.9a Using an LDR to make a light operated switch

Figure 13.9b Using a Schmitt trigger to improve the switching action

The transistor as a current amplifier

Let us consider again the graphs in *figure 13.4* (p. 85). In order to make the BC109 collector current change from zero to 20 mA, the base current has been increased from zero to 40 μA (0·04 mA). The ratio of the change in collector current (ΔI_c) to the change in base current (ΔI_b) is called the small signal forward current transfer ratio, or **current gain**, and is given the symbol h_{fe}. (There are no units as this is a ratio.)

$$\therefore \quad h_{fe} = \frac{\Delta I_c}{\Delta I_b} = \frac{20}{0{\cdot}04} = 500$$

The symbol h_{FE} is also used and is defined as follows:

$$h_{FE} = \frac{I_c}{I_b}$$

The student should use the two curves to show that $h_{fe} = h_{FE}$ only when the transfer characteristic is linear. However, for most practical purposes, we may assume that $h_{fe} = h_{FE}$.

The transistor as a voltage amplifier

Look again at the graph in *figure 13.8* (p. 87), the large change in collector voltage (about 6 V) has been brought about by a small change in base voltage (about 0·15 V). The ratio of the change in collector voltage to the change in base voltage is called the **voltage gain**.

$$\therefore \quad \text{voltage gain} = \frac{6}{0{\cdot}15} = 40$$

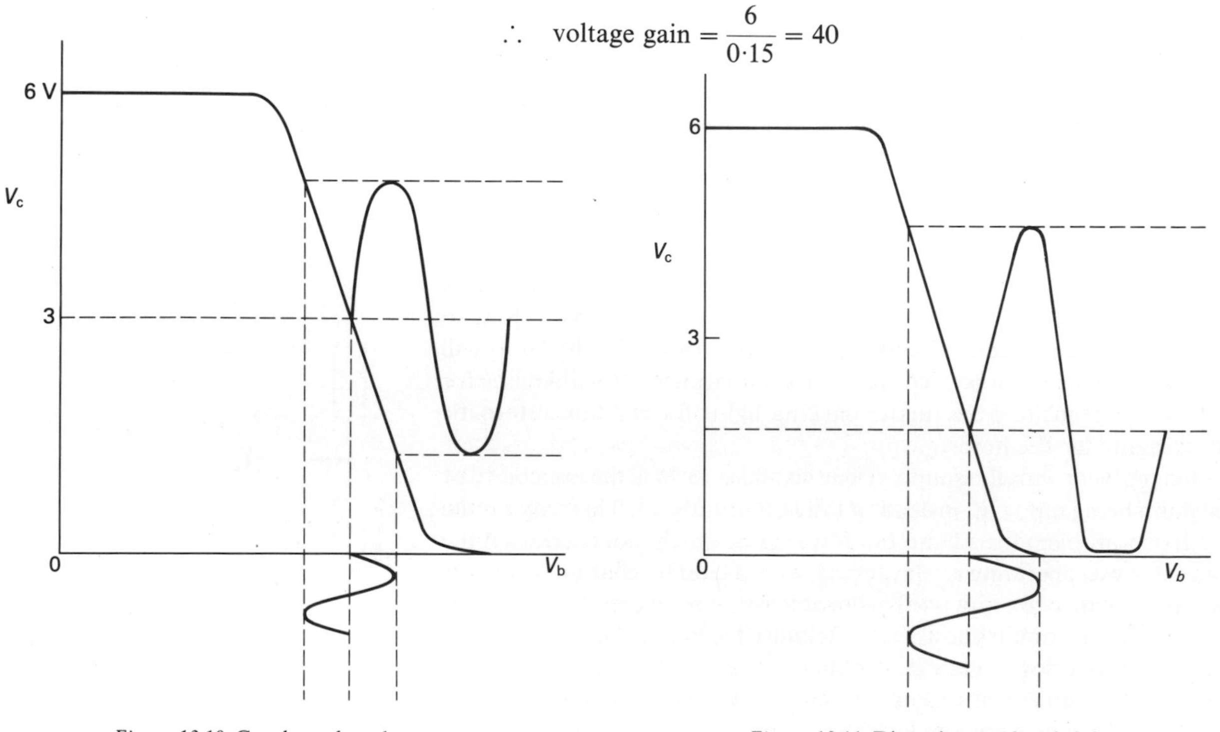

Figure 13.10 Graph to show how an a.c. signal can be amplified

Figure 13.11 Distortion produced if the mid-point is too low

The most important application for voltage amplification is for small a.c. signals, such as those obtained from microphones and pick-ups. *Figure 13.10* shows how an a.c. signal, correctly applied to the mid-point of the slope, can give an undistorted and amplified output. *Figure 13.11* shows the distortion of the wave which will occur if the fixed base voltage (called the **base bias**) is too high. *Figure 13.12* shows the distortion produced if the base bias is too low. Even if the base is correctly biased, it is still possible to produce a distorted output by applying too large an a.c. signal to the base (*figure 13.13*).

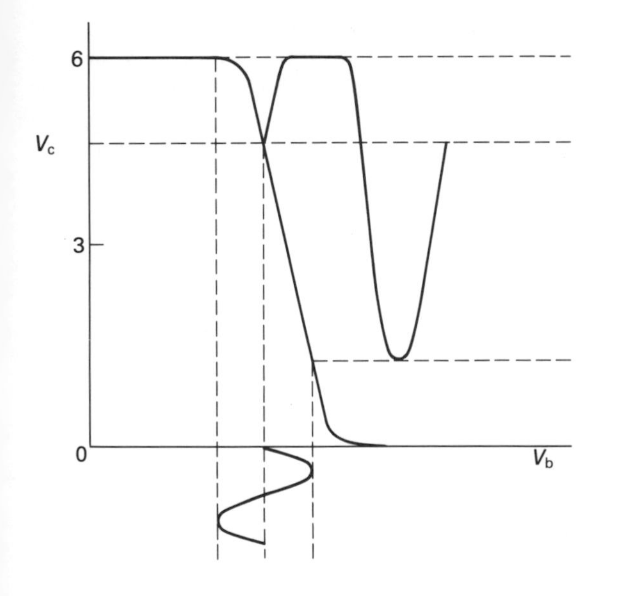

Figure 13.12
Distortion produced if the mid-point is too high

Figure 13.13
Distortion produced by too high an input signal

A circuit which can be used as a voltage amplifier is shown in *figure 13.14*. Notice the use of the capacitors C_1 and C_2 to lead the a.c. signal in and out, and at the same time prevent the d.c. at the base and collector interfering with other circuits. The base bias is set by means of the potential divider R_1 and R_2. The resistor R_L is called the **collector load**, and its action can be understood if you imagine it to be replaced by a piece of wire. The varying voltage at the base will make the collector **current** vary in sympathy, but we want a varying voltage at the output and this cannot happen because the collector **voltage** is held at +6 V by the wire to the power supply. If the collector current flows through a resistor, R_L, then any variations in current will produce a variation in voltage across this resistor, R_L. This gives the required output at the collector.

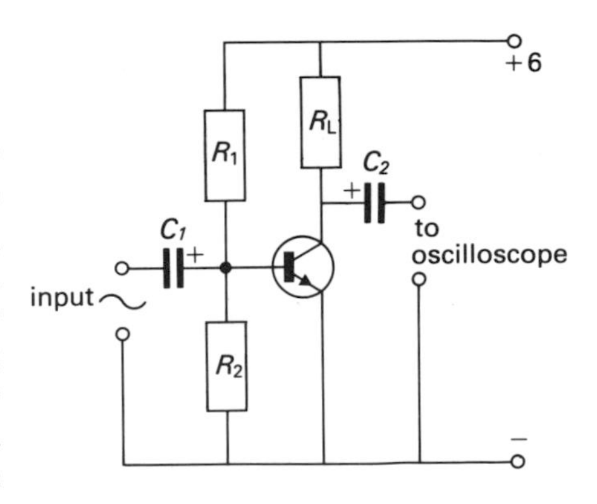

Figure 13.14 A complete single transistor amplifier

Calculating component values

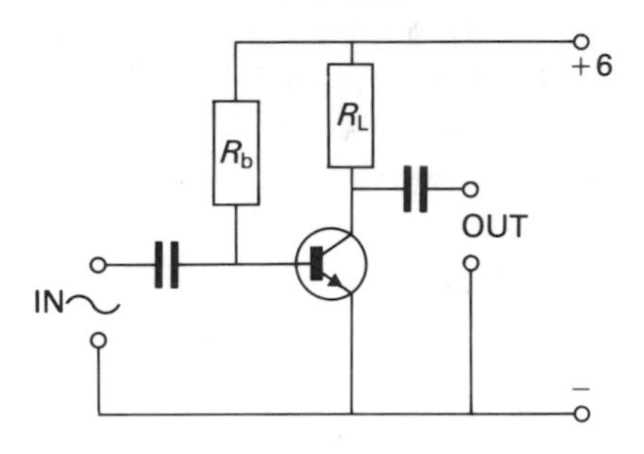

Figure 13.15

The simplest circuit for a common emitter amplifier is shown in *figure 13.15*. The collector voltage must be half the supply voltage (6 V) so that it can 'swing' by 3 V above and below this mid value without 'hitting' the top (6 V), or the bottom (zero volts). In the absence of any input signal, a suitable collector current would be 1 mA. This is called the **quiescent current**. To calculate the value of R_L:

$$V = IR$$

$$3 = 0{\cdot}001 \times R$$

$$R = 3\,\text{k}\Omega$$

If $h_{fe} = 400$ then the base current will be 1/400 of the collector current

$$\therefore \quad \text{base current} = \frac{0{\cdot}001}{400} = 2{\cdot}5\,\mu\text{A}$$

From *figure 13.8* we have seen the base voltage at the mid-point is about 0.67 V. This means R_b has +6 V at one end, and +0.67 V at the other end, so the potential difference across it is 5.33 V. To calculate the value of R_b: $V = IR$

$$5{\cdot}33 = 2{\cdot}5 \times 10^{-6} \times R$$

$$\therefore \quad R_b = \frac{5{\cdot}33}{2{\cdot}5 \times 10^{-6}} = 2{\cdot}13 \times 10^6\,\Omega$$

So the nearest preferred value (E12) is 2·2 MΩ.

However, *figure 13.15* is not a good circuit for two reasons:

a The calculated values are likely to be different if the transistor is changed, even if it is replaced by another of the same type, the reason for this being that the values of h_{fe} differ from one transistor to another.

b If the temperature rises, the collector current will rise owing to an increase in the **leakage current**. The leakage current is the current being carried by the thermally generated minority carriers (see unit eleven). This increase in collector current tends to produce a further rise in temperature and the effect escalates. This fault is called **thermal runaway**.

Preventing thermal runaway

A slightly improved circuit is shown in *figure 13.16*. With this circuit, there is some negative feedback to help prevent thermal runaway, but the value of R_b is still dependent upon the value of h_{fe}. In order to make the base voltage quite independent of h_{fe}, we use the potential divider R_1 and R_2 shown in *figure 13.17*. (Can you see why R_1 and R_2 must be high value resistors?). However, the negative feedback is now lost and hence, the protection against thermal runaway is lost.

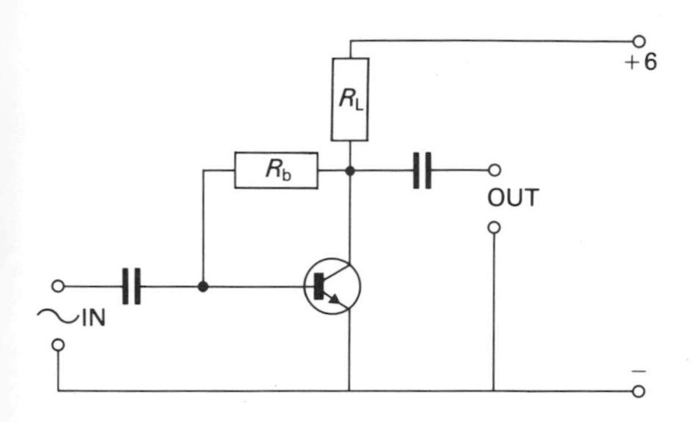

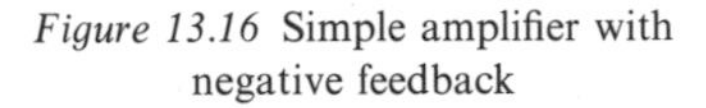
Figure 13.16 Simple amplifier with negative feedback

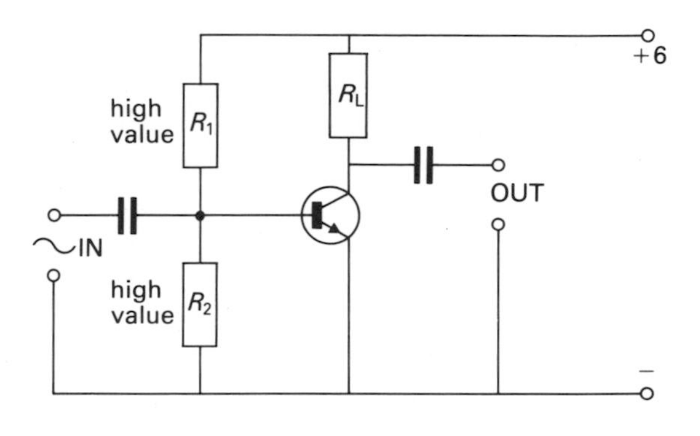

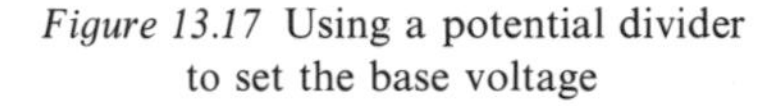
Figure 13.17 Using a potential divider to set the base voltage

Figure 13.18 shows the addition of the **emitter resistor**, R_E. If the temperature rises, the increased collector current gives an increased p.d. across R_E, because the current has to flow through it. If the p.d. across R_E rises, then the voltage at the emitter must also rise, since the other end of this resistor is fixed at zero volts. The base voltage is fixed by the potential divider, so the p.d. between the base and emitter falls due to the temperature rise. This fall in the base-emitter voltage tends to reduce the collector current, cancelling the effect of the temperature rise.

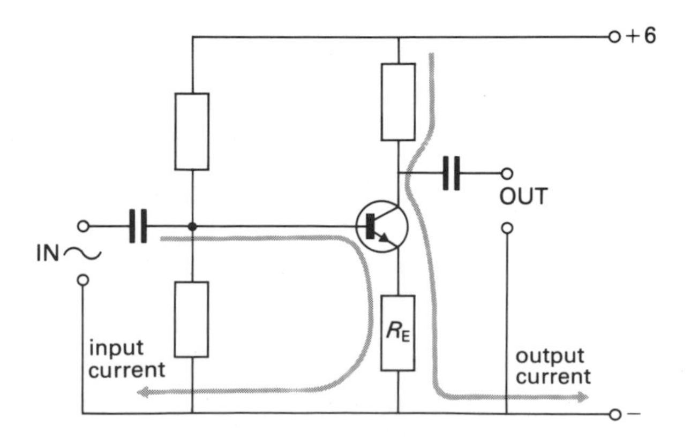

Figure 13.18 The flow of current through the emitter resistor

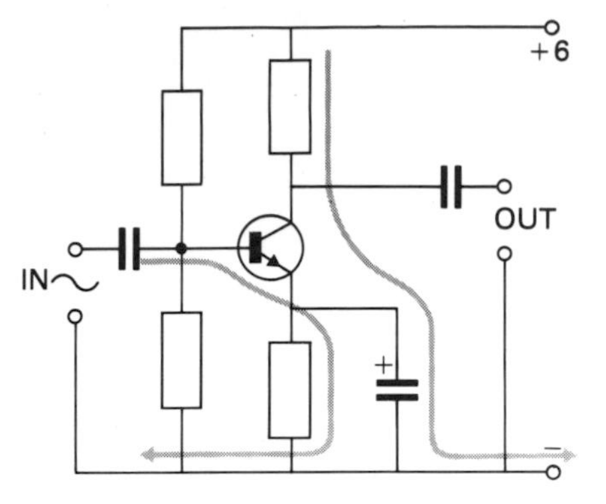

Figure 13.19 Using the emitter bypass capacitor to reduce negative feedback

The emitter bypass capacitor

The amplifier circuits we have been looking at are called **common emitter** circuits, because the emitter is common to the input and output. *Figure 13.18* shows how both input and output currents will flow through the emitter resistor. As the output signal waves of a transistor are exactly half a wave out of step with the input waves, the voltage developed across R_E, by the output current, tends to reduce the input voltage. The emitter resistor provides negative feedback, and if this feedback is not required, the voltage gain of the transistor would be increased by its removal. By placing a capacitor of low impedance across this resistor an alternative and easier path for the a.c. signals is provided. The emitter resistor now only carries d.c., and the negative feedback is removed without losing the temperature stability (*figure 13.19*).

The load line

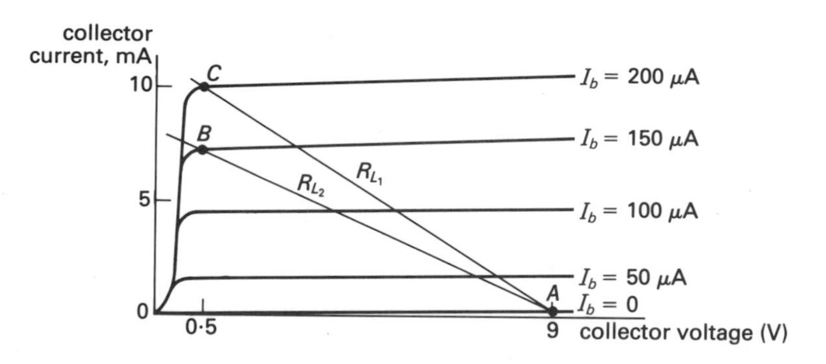

Figure 13.20 Plotting load lines

The usual method for calculating the value of the collector load resistor, R_L, is to make use of a set of curves known as the **common emitter output characteristic**. *Figure 13.20* shows a typical set, which can be explained quite simply. As soon as the collector voltage exceeds about 0·5 V, the value of the collector current is virtually independent of further increases in the collector voltage, so the resulting lines on the graph are horizontal. If the base voltage is zero there will be no base current ($I_b=0$), and so only the tiny leakage collector current will flow. There will be no voltage drop across R_L and so the collector voltage will be almost equal to the supply voltage (9 V in this case). This corresponds to the point *A* on the graph, and the transistor is said to be **cut off**. If the base voltage is made positive and increased, base current flows (say, 150 μA). This in turn produces a much larger collector current and there is a voltage drop across R_L. The collector voltage falls and follows the line *AB* (or *AC* for a base current of 200 μA). The lines AB and AC are called **load lines**, and at the points B and C the transistor is said to be **saturated**. The value for the collector load is calculated by applying Ohm's law as follows:

$$R_{L1} = \frac{\text{change in collector voltage}}{\text{change in collector current}}$$

$$\therefore \quad R_{L1} = \frac{9{\cdot}0 - 0{\cdot}5}{0{\cdot}01} = 850\,\Omega$$

The emitter follower

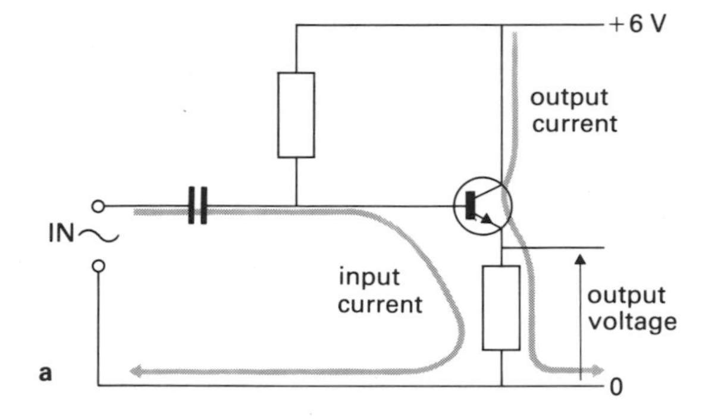

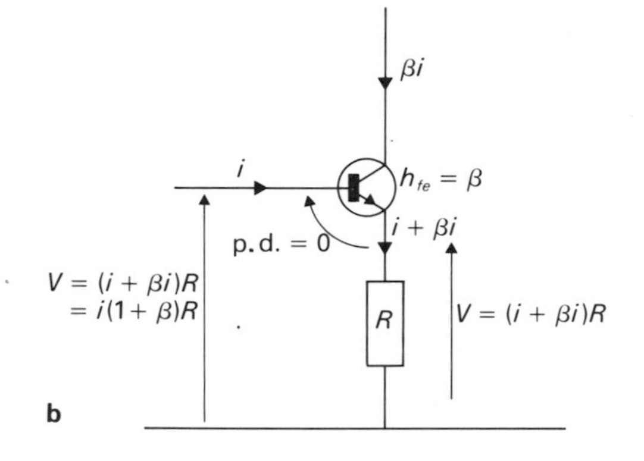

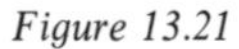

Figure 13.21

The emitter follower is not really a voltage amplifier, because the voltage gain will be slightly less than one. The circuit in *figure 13.21a* should be compared with *figure 13.15* (p. 91). *Figure 13.21a* shows that the collector load has gone, and the output must now be taken across the emitter resistor, through which the collector current flows. Since the entire output voltage appears across this resistor, which is still common to the input, there is 100% negative feedback. The reason for using such a strange circuit can be explained if we consider what has happened to the input resistance. If there was no negative feedback, the input signal voltage would drive a base current and the input resistance could be calculated. The negative feedback has reduced the input current, so the same signal now drives a smaller current and the input resistance is now effectively greater. *Figure 13.21b* shows that the input resistance is $(1+\beta)$ times the value of the emitter resistor, where β = current gain.

This circuit could be used as the first stage of a many stage amplifier in order to give it a high input resistance. This circuit is also called the **common collector**, because by connecting the collector directly to the power supply it is at **ground potential** for a.c. signal purposes. (Ground potential is the zero volt line to which all other voltages are compared.)

The common base amplifier

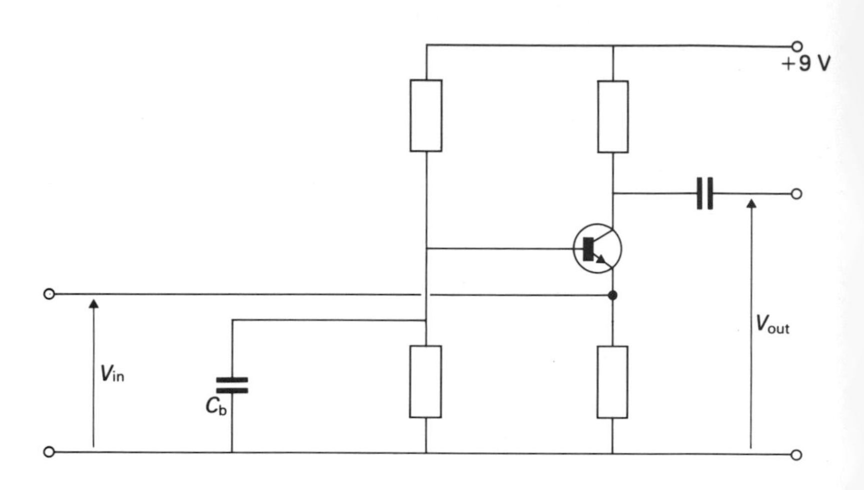

Figure 13.22

The third possible method of connecting a single transistor as an amplifier is called the **common base** circuit (*figure 13.22*). Notice first the similarity to the common emitter circuit in *figure 13.18*. It is more obvious that the differences are, firstly, the presence of the capacitor C_b and, secondly, the point at which the signal is fed in. The capacitor C_b should have a negligible impedance at the lowest operating frequency, and it is required here in order to keep the base at ground potential for the a.c. signal.

The chief characteristics of this circuit are its low input impedance (around 10 Ω), and its high output impedance. It is most useful for amplifying radio frequencies, because its high frequency performance is better than the common emitter circuit. It is also used as a means of supplying low voltage d.c. at a current which is almost independent of load. One of the objections to dry cells and cheap battery eliminators is the way their output voltage falls as the current they drive increases.

The differing electrical properties of a single stage transistor amplifier in its three possible modes are summarised in the following table:

	input impedance	**voltage gain**	**current gain**	**output impedance**
common emitter	medium	high	high	medium
common base	low	high	1	high
common collector	high	1	high	low

The constant current source

A constant current source is necessary, for example when recharging nickel-cadmium cells. A simple circuit for this purpose is shown in *figure 13.23*. The purpose of the Zener diode is to hold the base at a constant potential. As the p.d. between the base and emitter of a transistor is virtually zero, the p.d. across R will be 5 V in this case. This will drive a current of $5/R$ amps through R, and because the base current is very tiny, the collector current will be $5/R$ amps. Provided that the collector is always kept at a positive potential by the cell being charged, the collector current, and hence the charging current, remains constant.

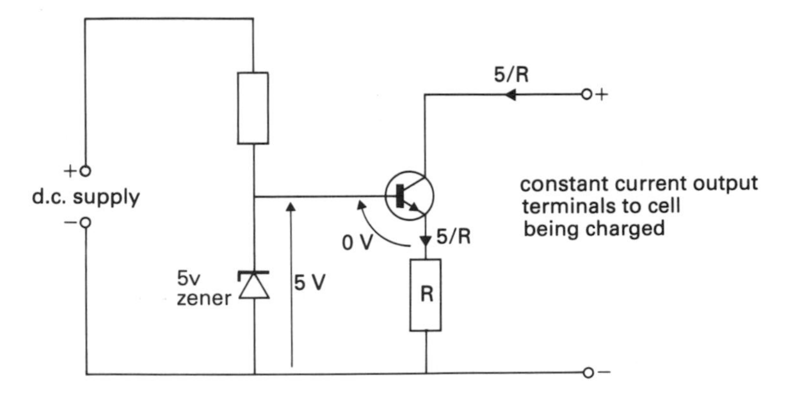

Figure 13.23 A constant current d.c. supply

d.c. amplifiers

It is sometimes necessary to amplify very low frequency a.c., or even d.c. signals. In this case, there must be no capacitors in the signal path. From the work done with the oscilloscope in unit six, it should be apparent that the Y amplifier of a CRO is a d.c. amplifier.

Testing a transistor

It is possible to use an ohm-meter to perform two simple tests on a transistor:

a to see if it is *n-p-n* or *p-n-p* type.

b to see if it has completely broken down.

If the base of an *n-p-n* transistor is made positive and the collector negative, this *p-n* junction will be on forward bias and its resistance will be low. If the emitter is made negative and the base positive, this junction will also have a low resistance. If the base is made negative, the collector-base and emitter-base junctions will be on reverse bias and so will have a high resistance. *Figure 13.24* illustrates this and also serves as a reminder that the positive terminal of most testmeters is the negative pole when the instrument is used as an ohm-meter.

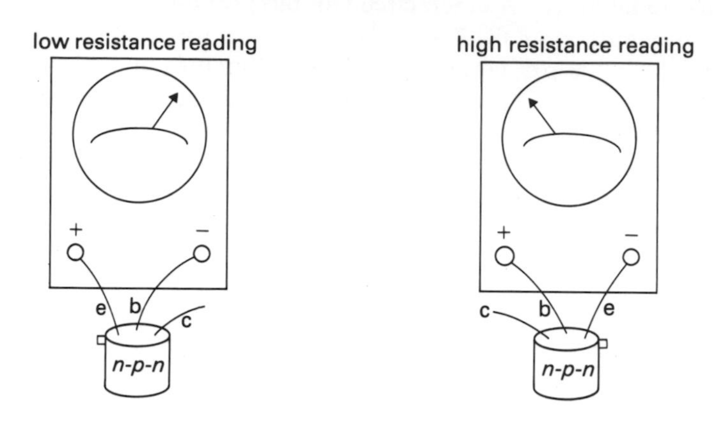

Figure 13.24
Using an ohm-meter to test an *n-p-n* transistor

If the two junctions, tested as above, do not have high and low resistance under the appropriate conditions the transistor has a major fault.

A similar test is used for *p-n-p* transistors, but with all signs reversed as shown in *figure 13.25*.

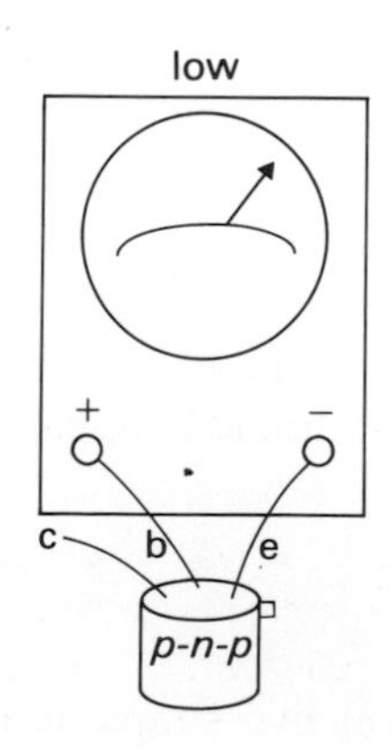

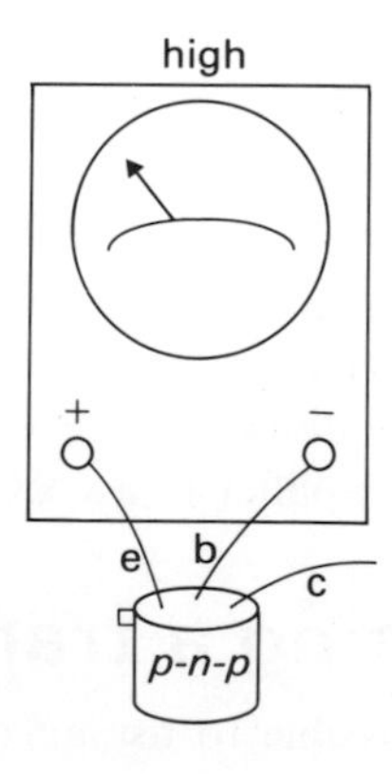

Figure 13.25 (*right*) Testing a *p-n-p* transistor

Figure 13.26 (*below*) Summary of ohm-meter tests

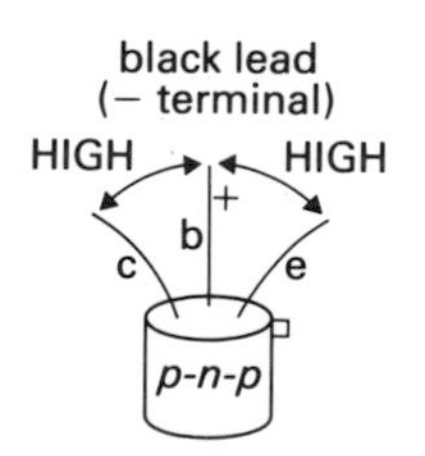

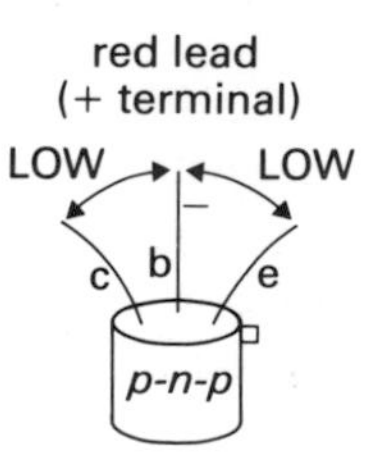

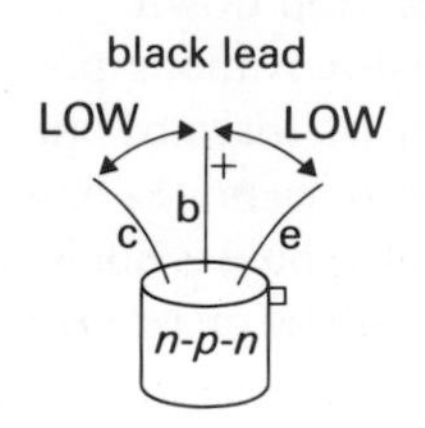

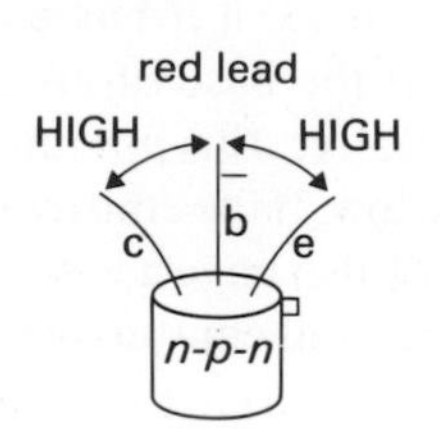

Other semiconductor devices

Thyristors

The **thyristor** is a four layer *p-n-p-n* device with three external connections (*figures 14.1–2*). In the absence of a voltage at the gate, the thyristor will allow only a tiny leakage current to flow from anode to cathode. By applying a small positive voltage to the gate, the thyristor is turned on, and it provides a conducting path of low resistance from anode to cathode. The thyristor will remain on, even if the gate voltage is removed or reversed in polarity. It will turn off again, only when the current from anode to cathode is externally reduced to zero. A thyristor will not conduct in the reverse direction, even with a voltage on the gate. The simple circuit shown in *figure 14.3* can be used to demonstrate these properties. It will be found that the lamp can be turned on by closing SW_1, and it will remain on until SW_2 is opened. If this experiment is repeated with a 6 V a.c. supply, SW_1 will turn the lamp on when closed, and off when opened, because the a.c. cycle reduces the anode–cathode current to zero one hundred times a second, on a 50Hz supply.

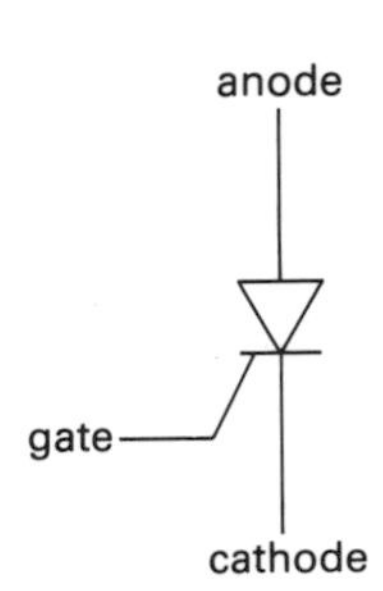

Figure 14.1 The thyristor symbol

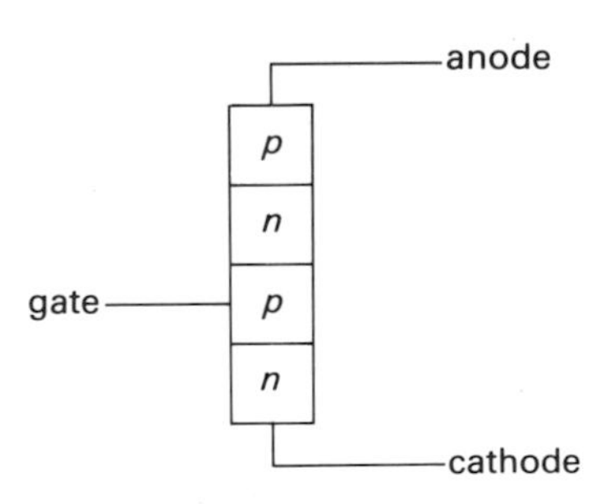

Figure 14.2 Schematic arrangement of junctions in a thyristor

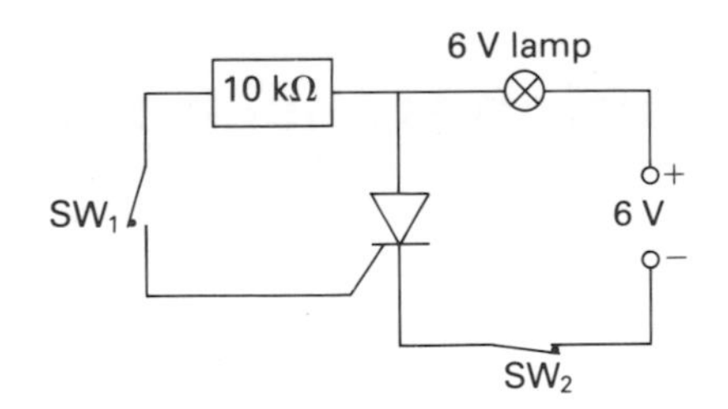

Figure 14.3 Circuit used to study the action of a thyristor

A thyristor could be used to turn on high current d.c. supplies by means of a small switch controlling the low gate current. It is not quite so easy, however, to turn the thyristor off, and a second thyristor with a second gate switch would be needed for this. The majority of thyristor applications are to be found in the control of a.c.

Triacs

The triac is a bi-directional thyristor which may be triggered into conduction in either direction by applying positive or negative pulses at the gate. The symbol for a triac is shown in *figure 14.4*. Notice that the current terminals are no longer labelled anode and cathode, because the

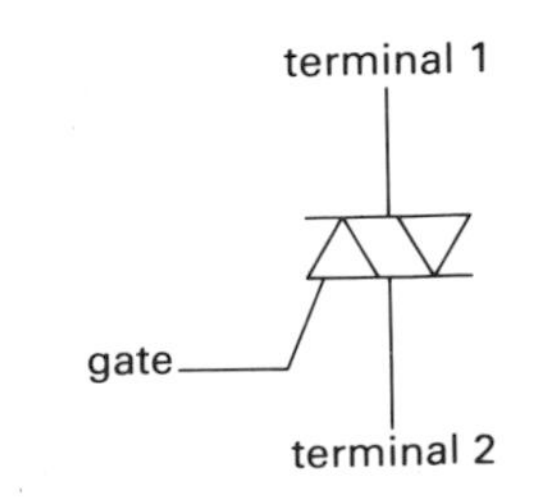

Figure 14.4 The triac symbol

triac can conduct in either direction. The triac is especially useful for switching a.c. because it conducts in both directions and therefore does not rectify, as would a simple thyristor.

Not only can a triac turn the a.c. on and off by means of a single low current switch, but it is possible for it to be turned on for only part of every a.c. cycle (*figure 14.5*). The point on each cycle where the triac is turned on can be adjusted by means of a simple circuit and this forms the basis of room light dimmers and electric drill speed controllers. It is necessary to include additional components to reduce the interference which such a device may cause with neighbouring radio and T.V. receivers.

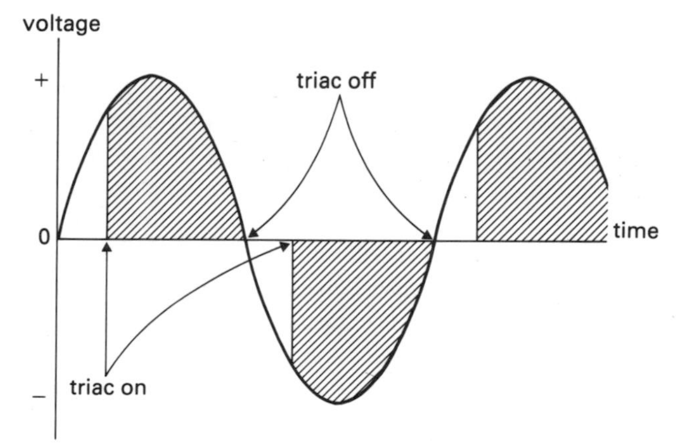

Figure 14.5 Voltage output from a triac when used as a dimmer

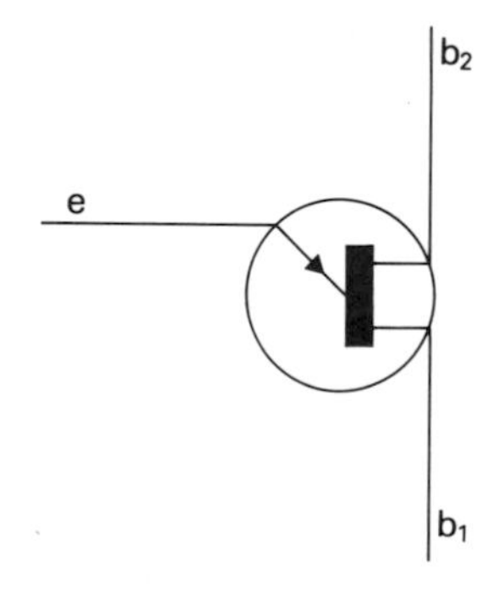

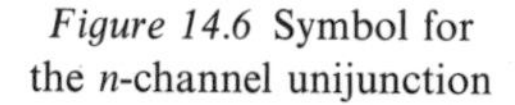

Figure 14.6 Symbol for the *n*-channel unijunction

The unijunction transistor

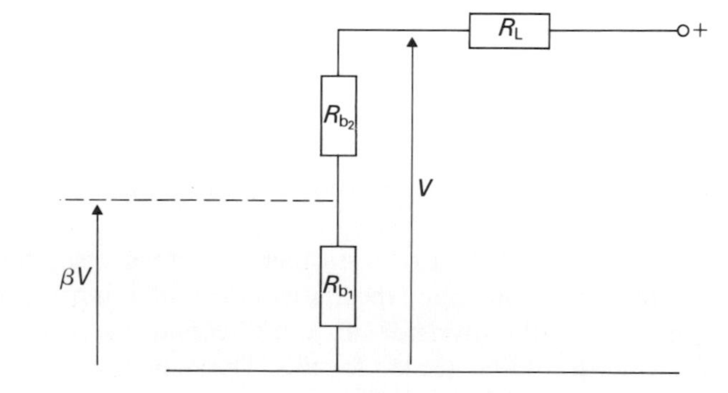

Figure 14.8
The unijunction acting as a potential divider

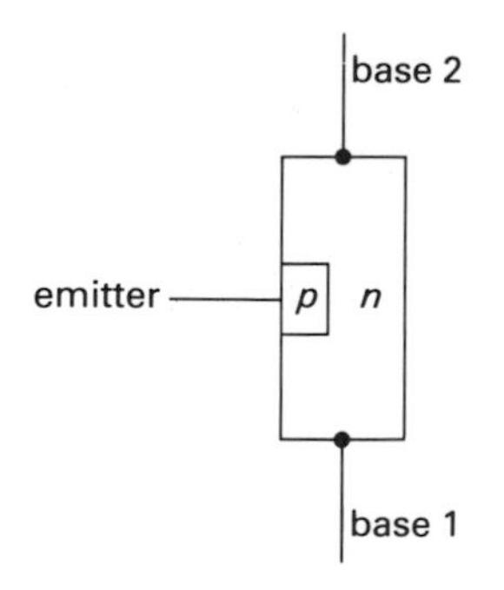

Figure 14.7 Simplified construction of the unijunction

As can be seen from *figure 14.6–7*, the unijunction is not a transistor, but a diode with two base connections. The resistance between b_2 and b_1, with zero volts on the emitter, is between 4 kΩ and 12 kΩ. The emitter is positioned closer to b_1 than to b_2, and, under these conditions the whole device acts as a potential divider (*figure 14.8*). Provided that the emitter voltage, V_e in *figure 14.9*, is less than a certain fraction of V, say βV, the emitter junction is reverse biased and no emitter current can flow. As soon as V_e rises above this critical value, the emitter junction becomes forward biased, resulting in a fall in resistance between emitter and b_1, and a steep rise in the emitter current.

A simple relaxation oscillator can be constructed as shown in *figure 14.10*. The capacitor C charges through the high value resistance R_1 until it reaches the potential which triggers the unijunction into conduction. The capacitor is then rapidly discharged through the low value resistor R_2. The capacitor then begins charging again and the cycle repeats at a rate which can be varied by altering the value of R_1 or C. The resulting oscillations are saw-tooth as shown in *figure 14.11*. This is a characteristic of relaxation oscillators.

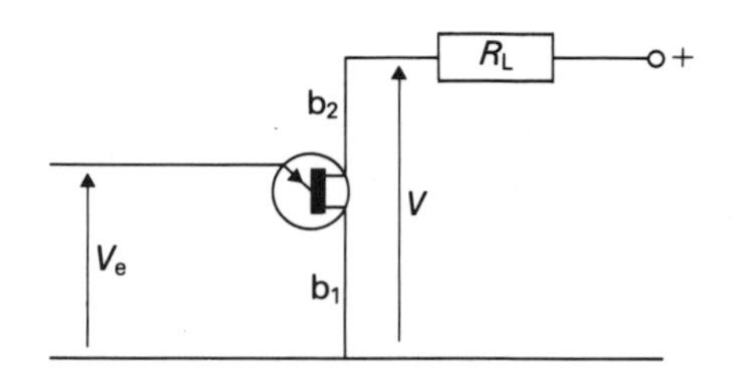

Figure 14.9
Voltages across a unijunction

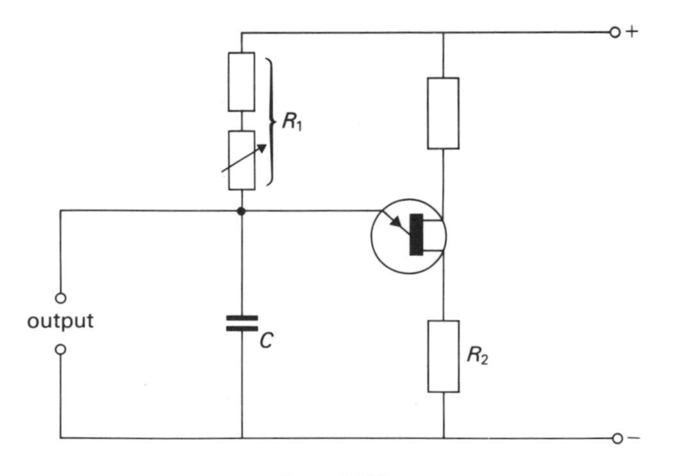

Figure 14.10
Circuit for a unijunction oscillator

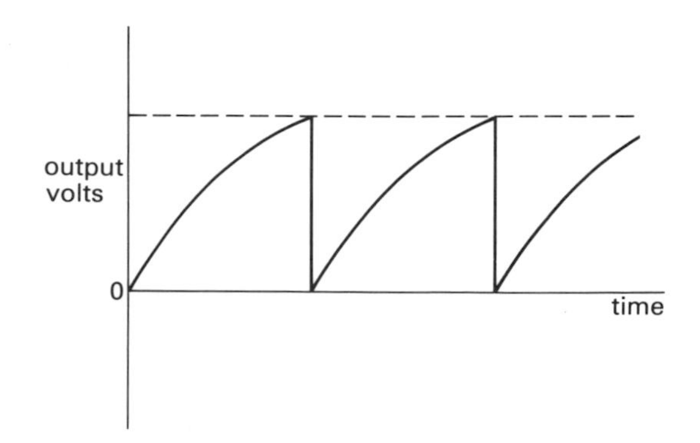

Figure 14.11 Sawtooth output voltage from a unijunction oscillator

The field effect transistor

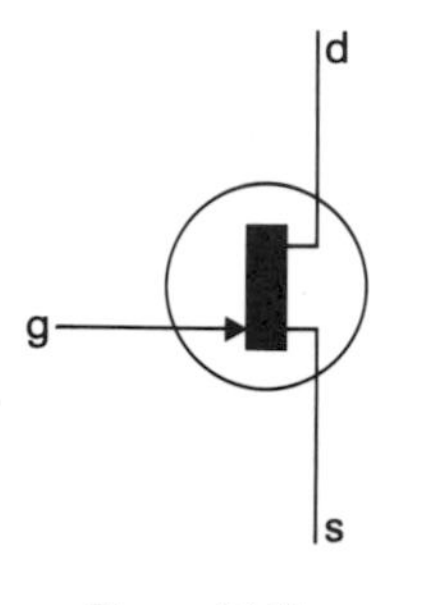

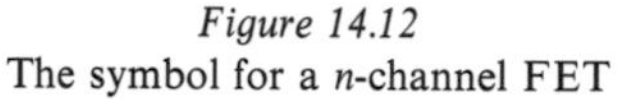
Figure 14.12
The symbol for a *n*-channel FET

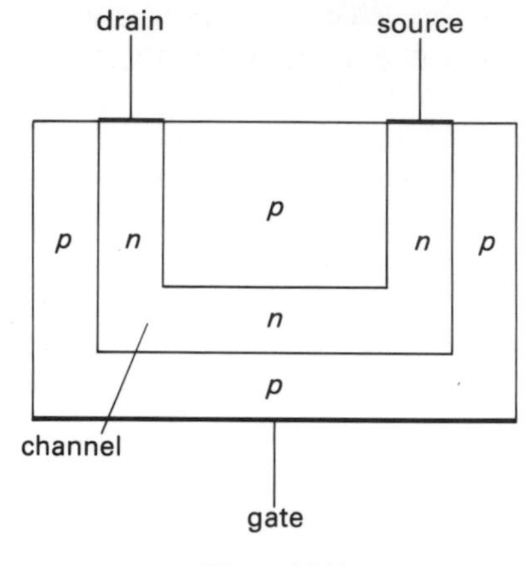

Figure 14.13
Construction of an FET in section

The field effect transistor (FET) has the construction and symbol as shown in *figures 14.12–13*. It is referred to as a **unipolar** device because only one type of charge carrier is involved (electrons in the case of the *n*-channel type). This is to contrast it with the more common **bipolar** transistor, in which both holes and electrons are involved. In an *n-p-n* bipolar transistor, holes injected into the base region provide a tiny base current, which controls the much larger electron current from emitter to collector. In an FET the current from source to drain is controlled by an electric field between the gate and the channel, hence the name **field effect** transistor.

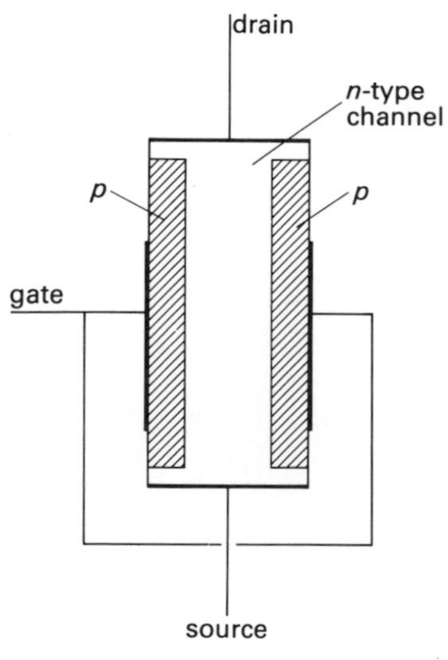
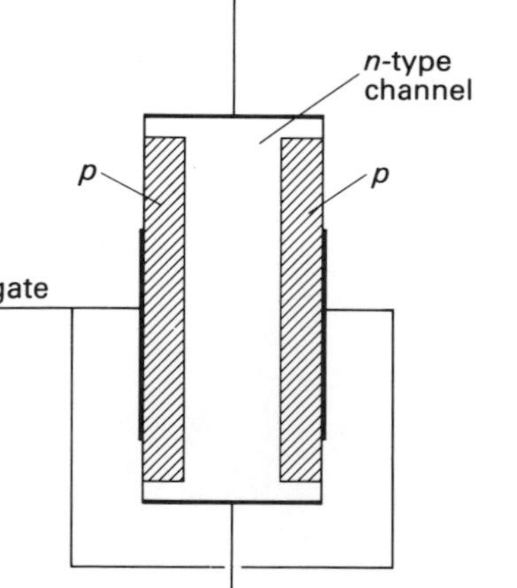

Figure 14.14
Construction of an FET shown simplified

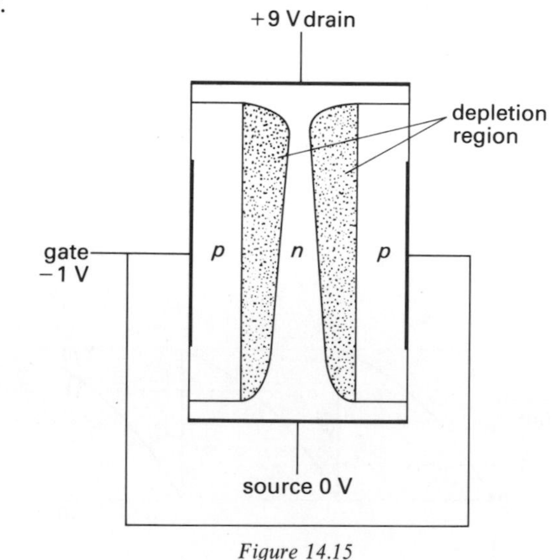

Figure 14.15

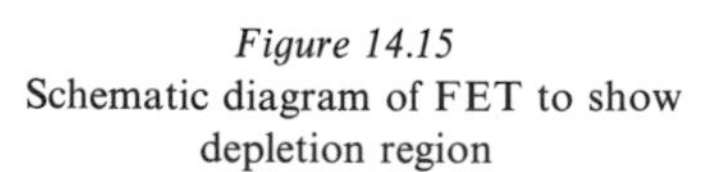
Figure 14.15
Schematic diagram of FET to show depletion region

A simplified form of construction is shown in *figure 14.14* where it can be seen to consist of a *p-n* junction between the *n*-type source/drain channel and the *p*-type gate. When this junction is on reverse bias there will be a depletion layer as shown in *figure 14.15*. This depletion layer is thickest at the drain end, because here the p.d. between drain and gate is greatest. The thicker the depletion region, the fewer charge carriers there are remaining in the channel, hence the higher the resistance of the channel between the source and drain. The FET behaves, therefore, like a voltage controlled resistor of value about 100 Ω with zero p.d. between the gate and the source. As the p.d. between the gate and the source (V_{gs}) becomes more negative, the resistance of the channel increases. When V_{gs} reaches the **pinch-off** voltage, the depletion layer will have almost completely closed the channel, and the source-to-drain resistance is now very high. Under these pinch-off conditions the drain current, I_d, is almost independent of the drain voltage, V_{ds}, as shown by the horizontal portions of the curves in *figure 14.16*. A typical circuit using an FET as a voltage amplifier is shown in *figure 14.17*. The a.c. signal causes a small swing in V_{gs} which in turn produces a much larger swing in V_{ds}.

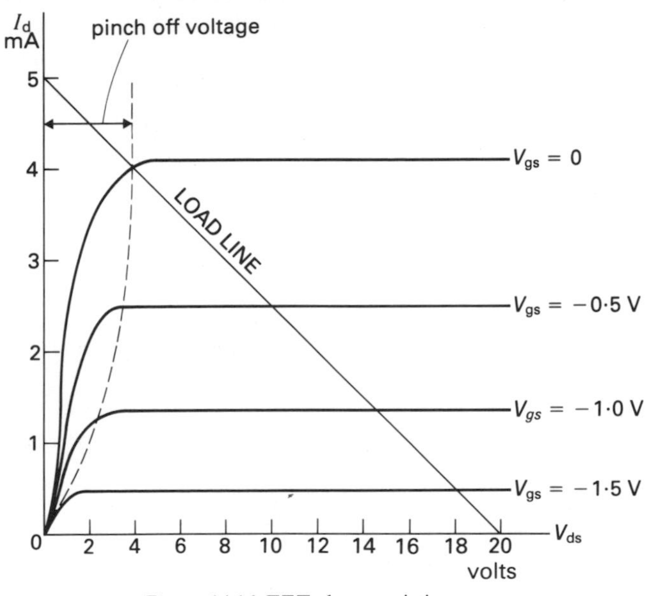

Figure 14.16 FET characteristic curves

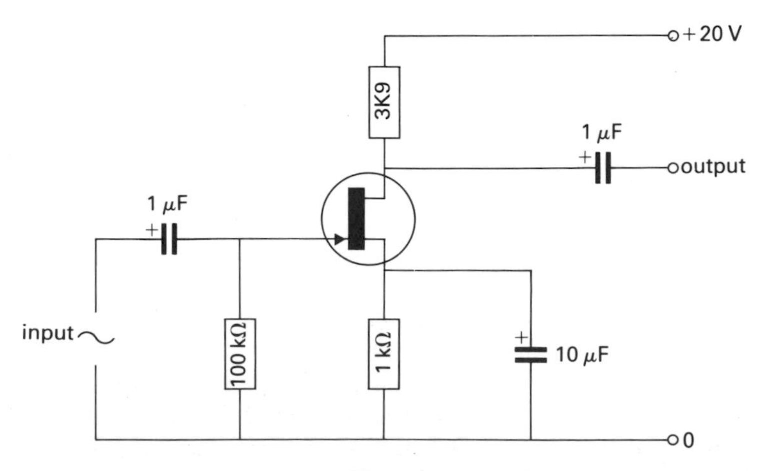

Figure 14.17 FET voltage amplifier

Unit 15

Electronic logic

Analogue and digital information

An electric current may carry information in a circuit in two quite different ways. In **analogue** circuits the information carried is related to the strength of the signal, and a continuous range of signal strengths is necessary to do this. In **digital** circuits only two types of signal exist, namely no current and some current. The strength of that current, provided it is not too high or too close to zero, is of little importance (*figure 15.1*). The two pieces of digital information carried are called logic 0 and logic 1.

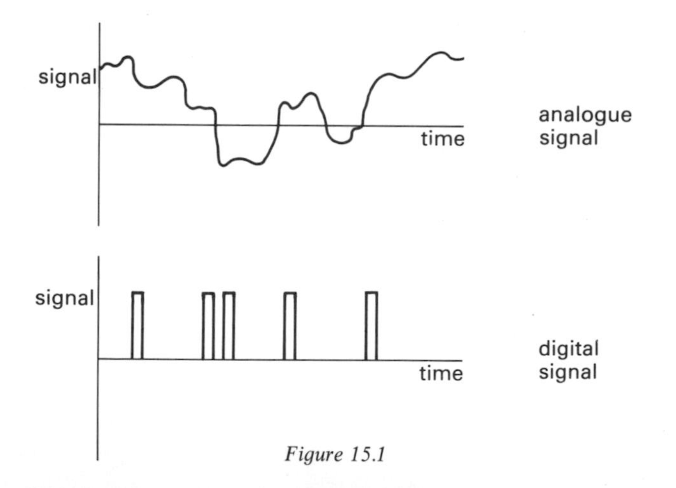

Figure 15.1

> logic 0 corresponds to any voltage.
> logic 1 corresponds to any other voltage, not too close to logic 0.

These two logic levels can be obtained simply by opening or closing a switch. Thus:

an open switch corresponds to level 0.
a closed switch corresponds to level 1.

The aim of this unit is to show how quite complicated information can be sent from one place to another by simply opening and closing switches. It is the branch of electronics called **digital logic**, and has its own special algebra.

Boolean algebra

The algebra of digital logic is quite different from ordinary algebra, and is known as Boolean algebra. We owe the origins of this to the Greek philosopher Aristotle who treated the thinking process as a chain of individual logical steps leading from some original statements to a final conclusion. The truth or falsity of the final conclusion depended upon the truth or falsity of all the individual steps. There could be no half truths. In 1847 an English philosopher called George Boole took Aristotle's ideas and based a logical algebra upon them. In 1938 Claude Shannon of the Massachusetts Institute of Technology showed that this algebra could be applied to circuits with switches in them.

The three basic Boolean algebraic operations, or functions, are called:

AND   OR   NOT

The AND function

The AND function arises when two or more switches are placed in series (*figure 15.2*). Current will only flow in the circuit when both *A* **and** *B* are closed. This is written in Boolean algebra as:

$$F = A . B$$

which is read: *F* equals *A* and *B*.

Figure 15.2 Two switches in series to illustrate the AND operation

Transistors may be used as the switches, and the device is then called a **logic gate**. The electrical symbol for a two-input AND gate is shown in *figure 15.3*. The use of **truth tables** is very helpful in order to describe fully the action of such a gate. *Figure 15.4* shows all the possible combinations of two switches, *A* and *B*. *Figure 15.5* shows the truth table for them; in it you see that 0 represents the open switch and 1 the closed switch. *Figure 15.6* shows the symbol for a three-input AND gate; see if you can draw up the truth table, before looking at *figure 15.7* on the next page.

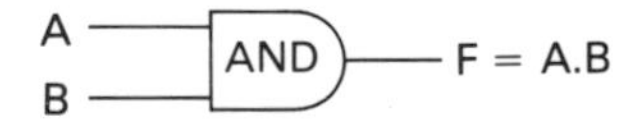

Figure 15.3 A two-input AND gate

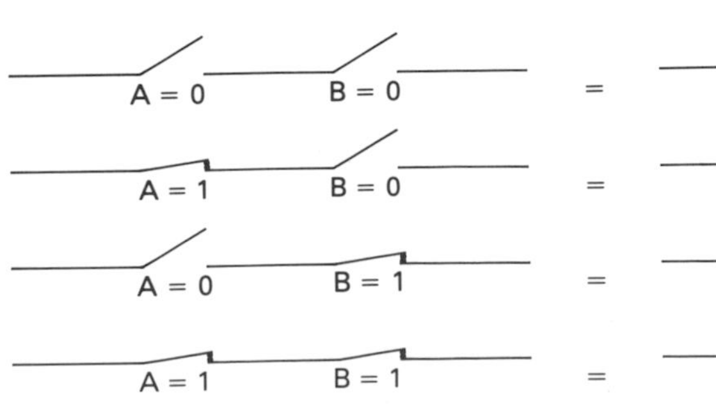

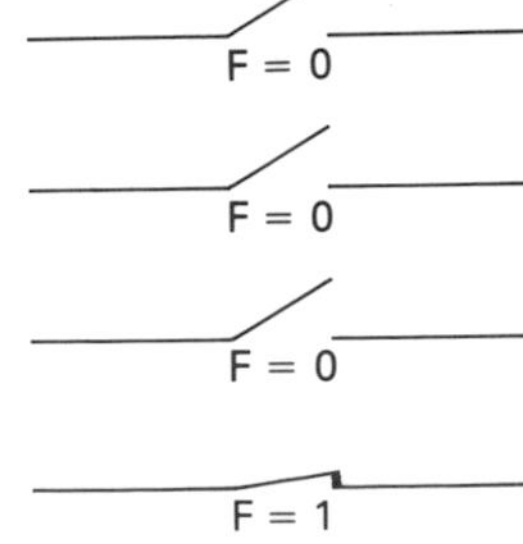

Figure 15.4 The possible combinations of two series switches

A	B	F = A.B
0	0	0
0	1	0
1	0	0
1	1	1

Figure 15.5
Truth table for a two-input AND gate

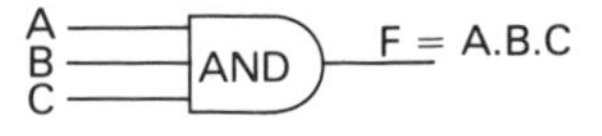

Figure 15.6 Symbol for three-input AND gate

A	B	C	F
0	0	0	0
0	0	1	0
0	1	0	0
0	1	1	0
1	0	0	0
1	0	1	0
1	1	0	0
1	1	1	1

Figure 15.7 Truth table for a three-input AND gate

In a package the size of the photograph below it is possible to contain 4 two-input AND gates (or 3 three-input AND, or 2 four-input AND). This package is called a fourteen pin dual in line (DIL) case. It has 14 leads, 12 of which are used for the gates, and 2 of which are used for the power supply of 5 V d.c.

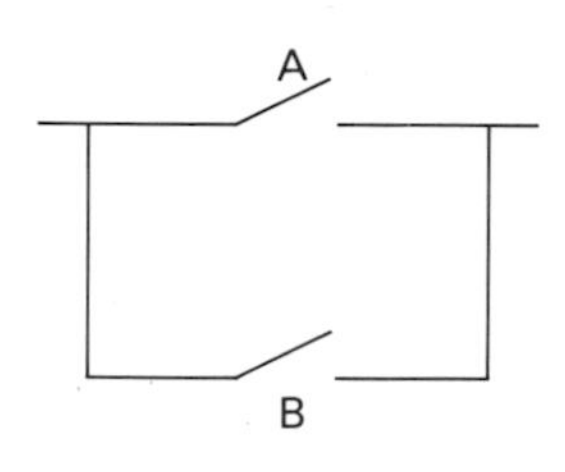

Figure 15.8 Switches in parallel to illustrate the OR operation

The OR function

This function arises when two or more switches are placed in parallel (*figure 15.8*), current will flow in this circuit when either *A* **or** *B* are closed. Current will also flow when both *A* **and** *B* are closed, so this OR function includes the AND function, hence it is called the inclusive OR. In Boolean algebra it is written:

$$F = A + B$$

which is read: *F* equals *A* or *B*

The electrical symbol for a two-input inclusive OR gate with its truth table is shown in *figure 15.9*. See if you can draw a truth table for a three-input inclusive OR.

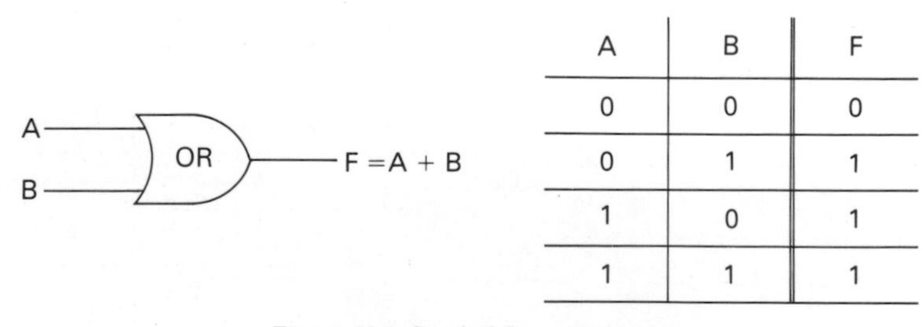

A	B	F
0	0	0
0	1	1
1	0	1
1	1	1

Figure 15.9 Symbol for a two-input OR gate and its truth table

The exclusive OR function

We have seen that the inclusive OR includes the AND function, so the exclusive OR must exclude the AND. The symbol for a two-input exclusive OR gate with its truth table is shown in *figure 15.10*.

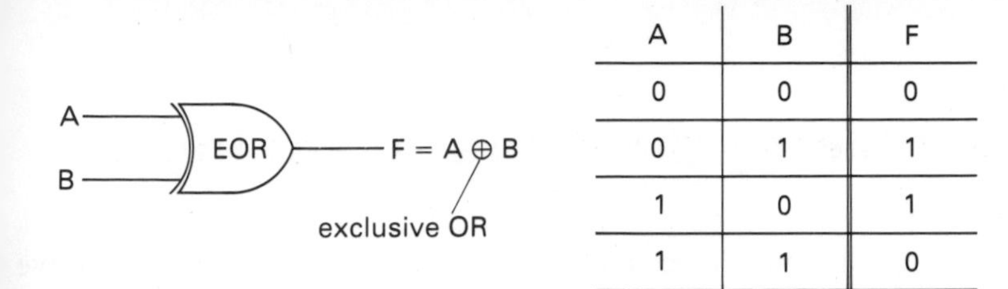

A	B	F
0	0	0
0	1	1
1	0	1
1	1	0

Figure 15.10 Symbol for two-input exclusive OR gate and its truth table

The invert, negate or NOT function

This operation simply involves changing a logic 0 to a logic 1 and a logic 1 to a logic 0. Its symbol is a bar over the letter, $\bar{A}$ which is read NOT A. The electrical symbol for an invert gate with its truth table is shown in *figure 15.11*.

When the base of a transistor amplifier is at zero potential its collector potential is high; when the base is high the collector is low. A single transistor does, therefore, act as an inverter.

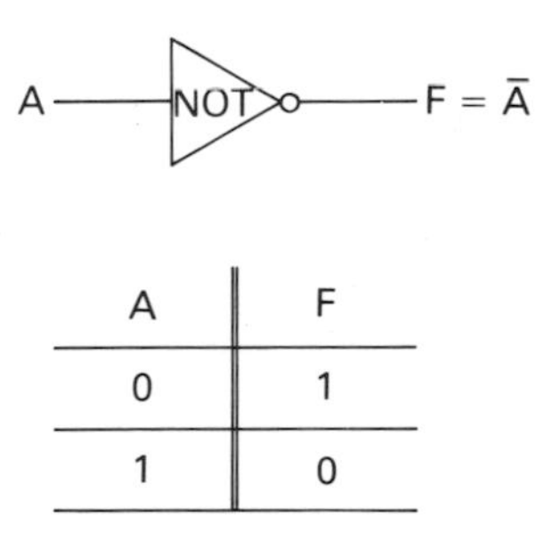

A	F
0	1
1	0

Figure 15.11 Symbol for an invert gate and its truth table

The NAND and NOR gates

When a transistor is being used to construct an AND gate, it also inverts the output. The resulting AND followed by a NOT is called a NAND (NOT AND) gate. Its electrical symbol and truth table shown in *figure 15.12*. Similarly an inclusive OR followed by a NOT is called a NOR (*figure 15.13*).

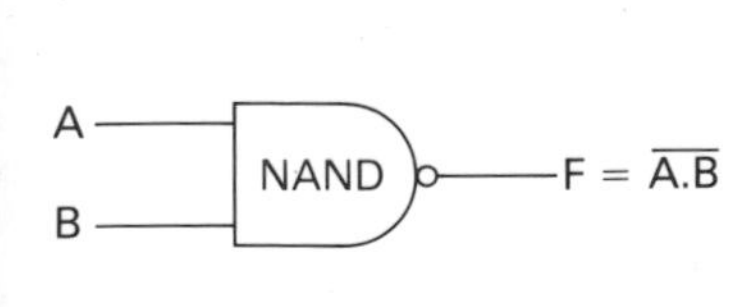

A	B	A.B	$\overline{A.B}$
0	0	0	1
0	1	0	1
1	0	0	1
1	1	1	0

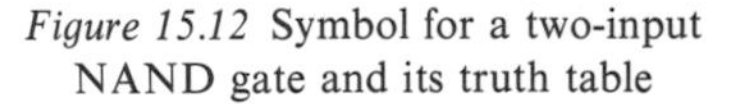

Figure 15.12 Symbol for a two-input NAND gate and its truth table

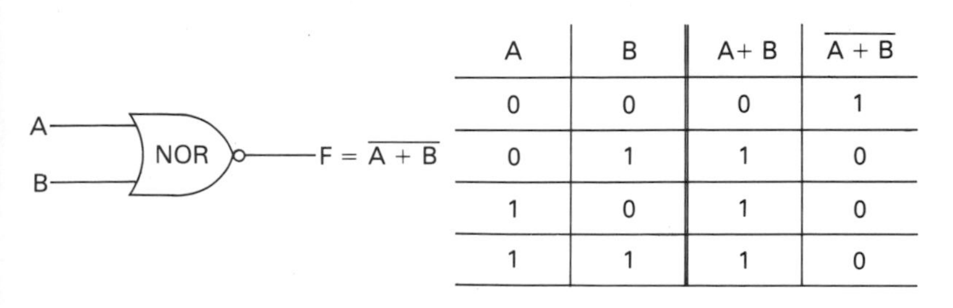

A	B	A+ B	$\overline{A + B}$
0	0	0	1
0	1	1	0
1	0	1	0
1	1	1	0

Figure 15.13 The two-input NOR gate and truth table

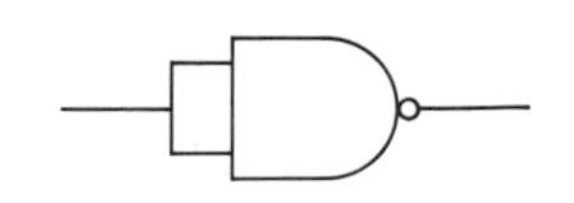

Figure 15.14 Wiring a two-input NAND gate to act as an inverter

All of these gates are readily available in standard fourteen-pin DIL packages, but it is possible to turn a NAND gate into a NOT gate, as shown in *figure 15.14*.

Combining the operation of logic gates

The study of digital logic can become quite fascinating once you have learned how to find the combined effect of several logic gates. There are two ways of doing this, namely to wire the circuit up on a logic tutor or to learn some of the rules of Boolean algebra and work the results out on paper.

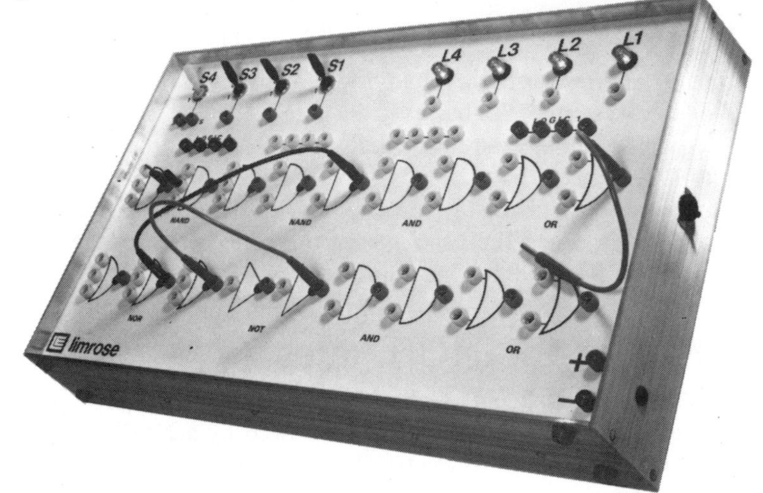

A combinational logic tutor

de Morgan's theorem

One of the most useful rules of Boolean algebra is de Morgan's theorem:

In order to negate (invert) a Boolean expression:

a negate any term which is not already individually negated and remove the negation from any term which is already individually negated.

b interchange AND and OR symbols between the terms.

The main use of de Morgan's theorem is to simplify expressions with several negations in them. So, for example, to simplify $\overline{\bar{A}.B}$, which is the output of the circuit *figure 15.15*, proceed as follows. There are two terms in this expression, A and B, so to carry out step **a,** remove the negation from $\bar{A}$ and place one on B. Between these two terms there is an AND, so to carry out step **b**, change the AND to an OR.

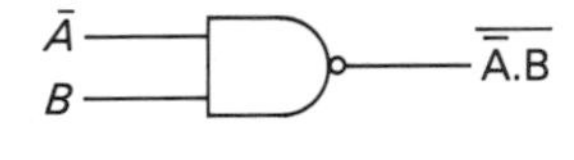

Figure 15.15

$$\therefore \quad \overline{\bar{A}.B} = A+\bar{B}$$

Replacing AND gates with NAND

The circuit shown in *figure 15.16* will produce an output of 1 whenever $A.\mathrm{B} = 1$, or when $C.\mathrm{D} = 1$. The same result can be obtained by using NAND gates, as in *figure 15.17*.

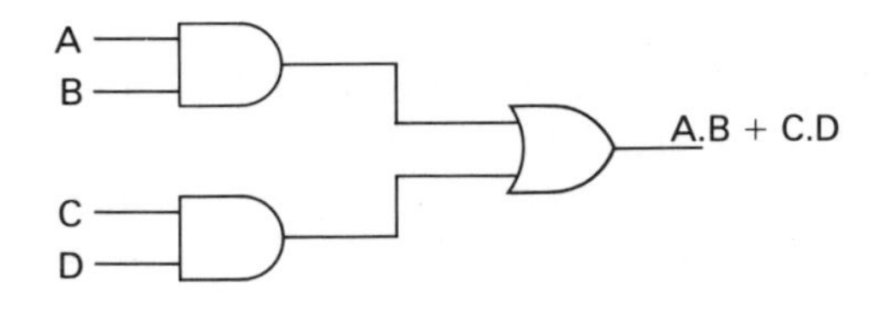

Figure 15.16

See if you can apply de Morgan's theorem to prove that

$$\overline{\overline{A.B}.\overline{C.D}} = A.B + C.D$$

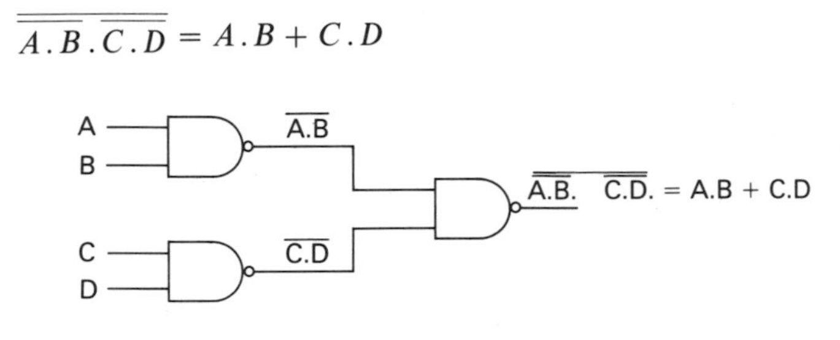

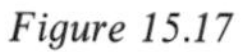

Figure 15.17

The comparator

Figure 15.18 shows the circuit of a simple comparator, or equality detector, the purpose of which is to compare the two input signals *A* and *B* and generate an output 1 when *A* = *B*. The truth table for this circuit is shown in *figure 15.19*.

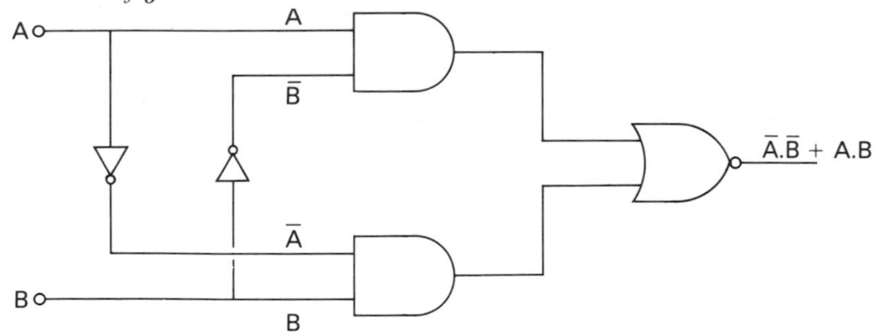

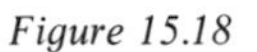

Figure 15.18

A	B	output	
0	0	1	($\overline{A}.\overline{B}$)
0	1	0	
1	0	0	
1	1	1	(A.B)

Figure 15.19

Practical gate circuits

When a manufacturer supplies a logic gate, the internal construction of that gate is of little real importance, provided you know how to use it and what its limitations are. It is instructive, however, to consider the construction of one simple gate using just diodes and resistors. *Figure 15.20* shows a simple two-input AND gate. In this circuit, if either *A* or *B* is at 0 V, then the associated diode will be on forward bias and its very low resistance will effectively reduce the potential at *X* to nearly zero. When both the inputs are 6 V, and the diodes are placed on reverse bias, there will be only high resistance paths between *X* and ground, so almost no current will flow through the 10 kΩ resistor, and the potential at *X* will rise to nearly 6 V. So, the output will be high when both the inputs are high.

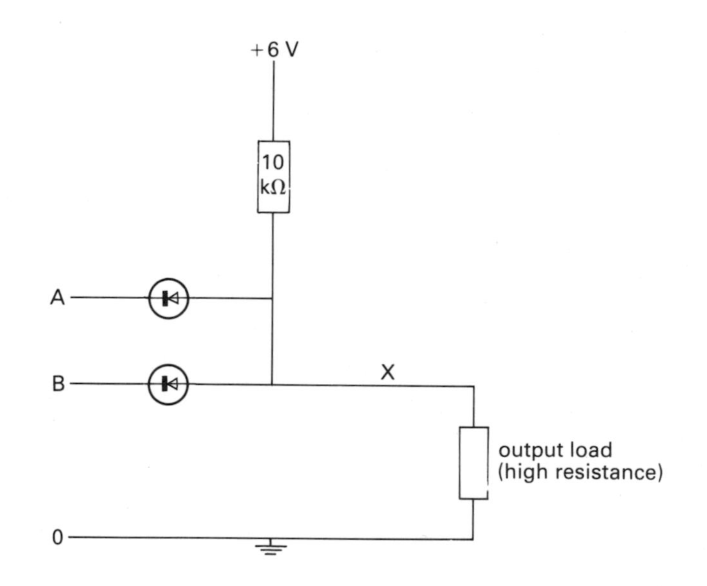

Figure 15.20 A simple diode AND gate

Multivibrators

Unit 16

A multivibrator is a type of oscillator which produces a square wave rather than a sine wave output. A square wave voltage varies and behaves as if it was made up from a very large number of frequencies, each of which is a multiple of the square wave frequency. For this reason the device which produces it has come to be called a multivibrator.

Types of multivibrator

Multivibrators fall into three groups:

a bistable, with two stable states.
b monostable, with one stable state.
c astable, with no stable state.

They are all based upon the action of the transistor as a switch in a circuit as shown in *figure 16.1*. The input wave switches the transistor alternately on and off, so producing a square wave output. If two of these switches are joined together and the output of the second is taken to drive the input of the first, the circuit in *figure 16.2* is obtained.

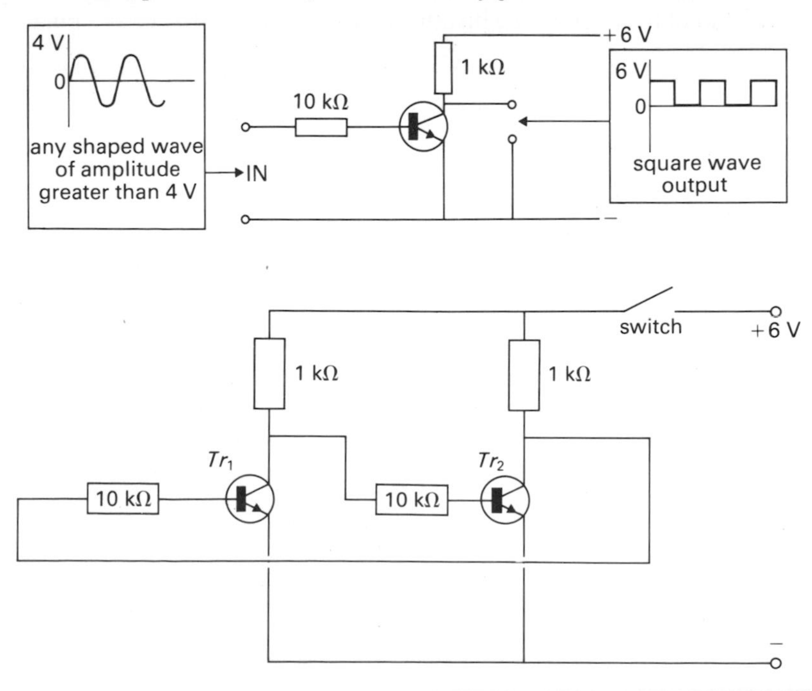

Figure 16.1 Using the transistor as a switch

Figure 16.2 Two transistor switches with feedback

When the switch is closed, the currents flowing through the two transistors are not equal, owing to slight differences in the resistors and transistors. One transistor will have a greater collector current than the other, let us say Tr_2 current is the greater. As Tr_2 collector current rises, so the collector voltage falls. Tr_1 still has a high base voltage and so is still conducting. As Tr_2 collector voltage falls, so does Tr_1 base voltage, and this begins to turn Tr_1 off. When Tr_1 is turned off its collector voltage will rise and quickly drive Tr_2 to saturation (full on). Tr_1 is now off, and Tr_2 is on. This state will remain as long as the 6 V supply is connected; it is called a **stable state.**

There are several ways of reversing the stable states of the two transistors:

a briefly connect the collector of the **off** transistor to ground (zero volts).

b briefly connect the collector of the **on** transistor to +6 V.

c briefly connect the base of the **off** transistor, via a resistor, to +6 V.

d briefly connect the base of the **on** transistor, via a resistor, to ground.

Of these four methods, the last is generally used.

Bistable multivibrators

The complete circuit of a **bistable** is shown in *figures 16.3–6*. This circuit is also called a **flip-flop** because each transistor flips from one state to the other and back again. In *figure 16.3* a rising edge of the square wave input has no effect; it just puts D_1 on reverse bias along with D_2. In *figure 16.4*, the falling edge of the input makes D_1 conduct; C_1 charges and positive feedback operates as shown until Tr_1 is off and Tr_2 on. In *figure 16.5*, C_1 discharges as the square wave goes positive. In *figure 16.6* the next falling edge makes D_2 conduct and the same process occurs with the other half of the circuit. The bistable will be used later to act as a binary divider.

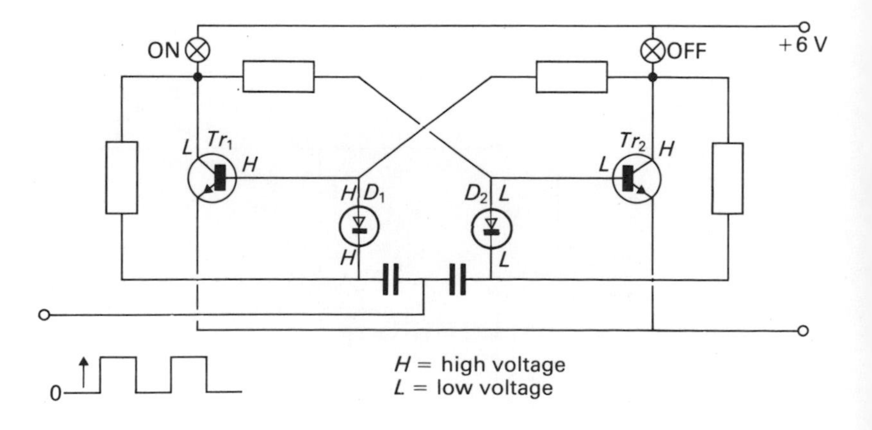

Figure 16.3
The bistable in its first stable state

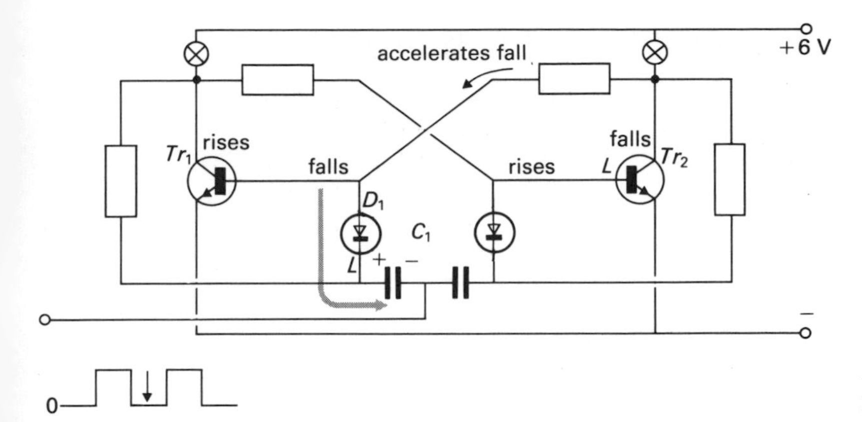

Figure 16.4 The bistable receives a trigger pulse

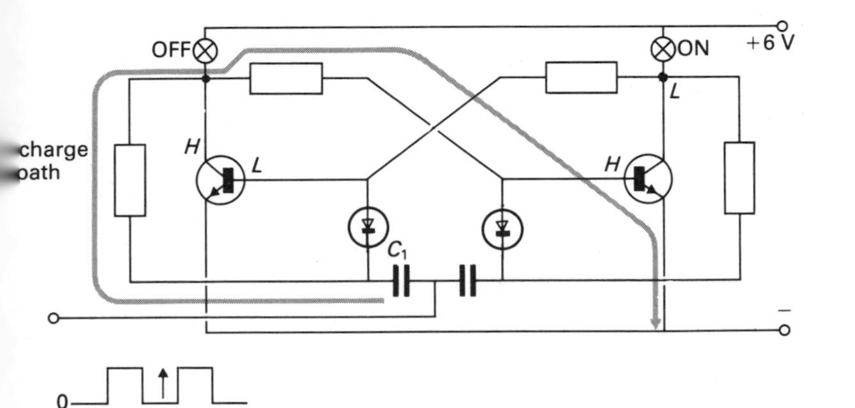

Figure 16.5 The bistable changes to its second stable state

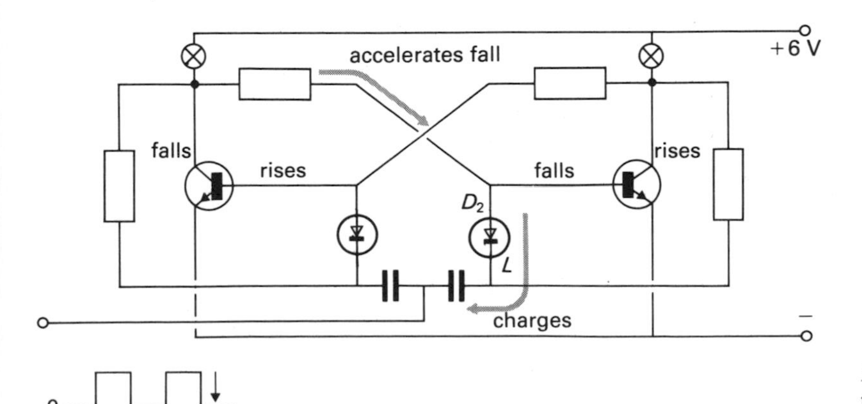

Figure 16.6 The next change of state is triggered

Monostable multivibrator

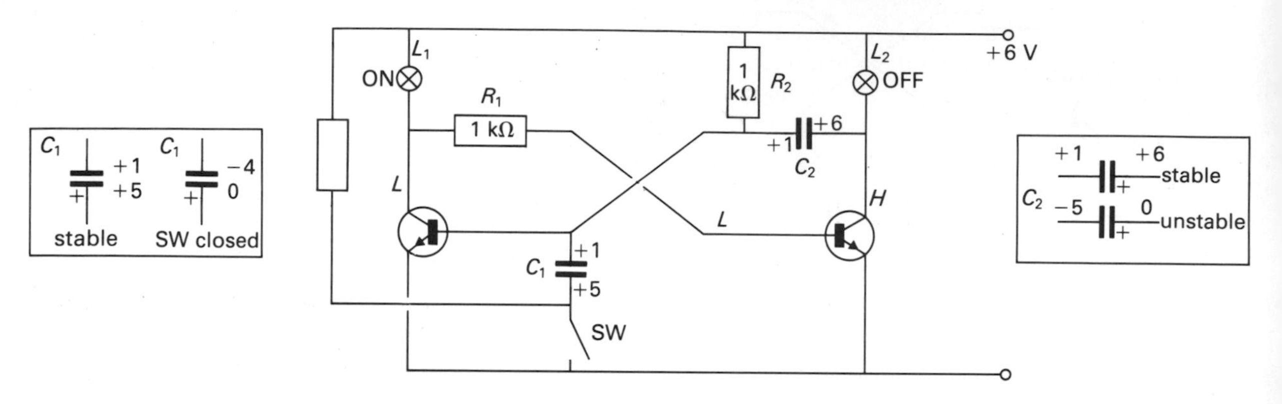

Figure 16.7 (above) The monostable in its stable state

Figure 16.8 (below) Changing the potential on one side of a capacitor

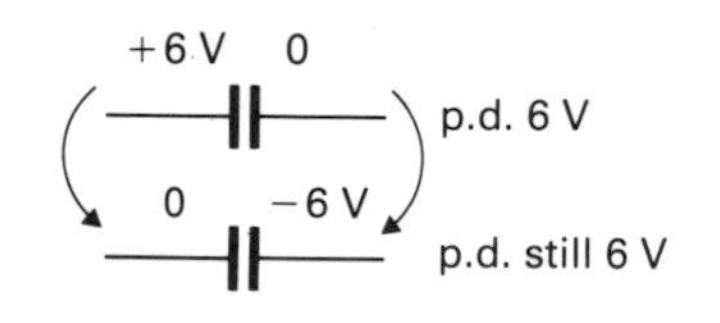

A circuit to show the action of a **monostable**, or one-shot multivibrator, is shown in *figure 16.7*. To understand the action of both the monostable and the astable circuits, it is necessary to know what happens when the voltage on one side of a capacitor is changed. This is illustrated in *figure 16.8*, the capacitor is charged and has a p.d. of 6 V across its plates. If the +6 V plate suddenly falls to 0 V, the capacitor does not lose its charge or change its p.d., so the plate which was at zero must fall to −6 V. Now look at *figure 16.9*, where C_1 is charged as shown. When the switch is closed, the plate which was at about +1 V now drops to −4 V. This turns Tr_1 off and the collector of Tr_1 rises to nearly +6 V, which turns Tr_2 on via R_1, and L_2 lights up as shown in *figure 16.9*. The capacitor C_2 now charges via R_2 and Tr_2 as shown in *figure 16.10*. The length of time the lamp, L_2, remains lit depends upon the time constant $C_2 \times R_2$.

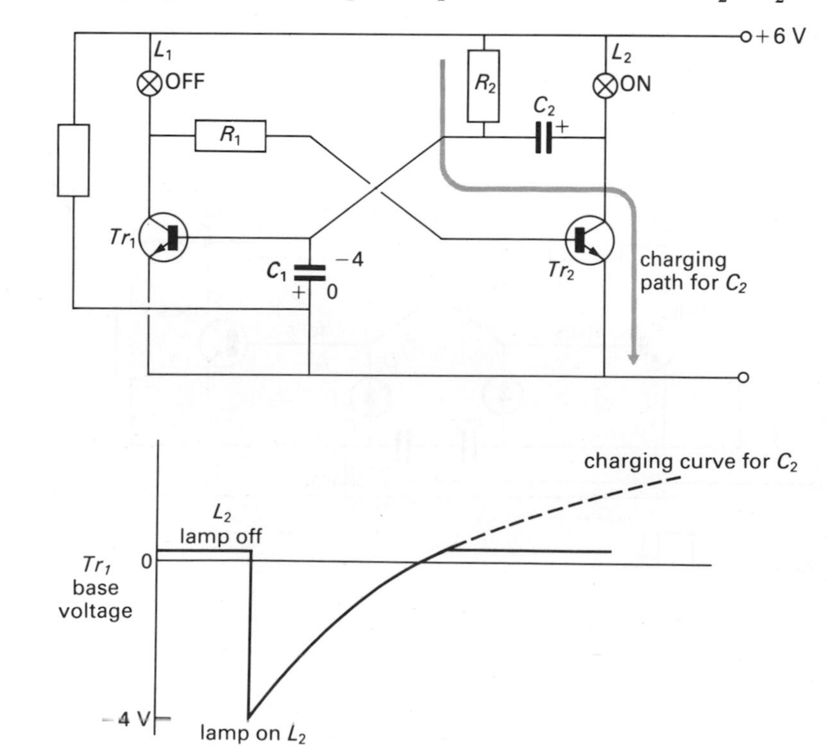

Figure 16.9 The monostable changes its state briefly

Figure 16.10 Charging curve for capacitor C_2

Astable multivibrators

The astable, or free-running multivibrator, is shown in *figures 16.11–14*. If you have understood the action of the monostable, you should be able to follow the sequence of diagrams showing how C_1 and C_2 alternately charge and discharge through Tr_2 and Tr_1. A dual trace oscilloscope, connected to the base and collector of Tr_1 or Tr_2 can be used to see how the voltages here vary with time. The results should look like *figure 16.15*.

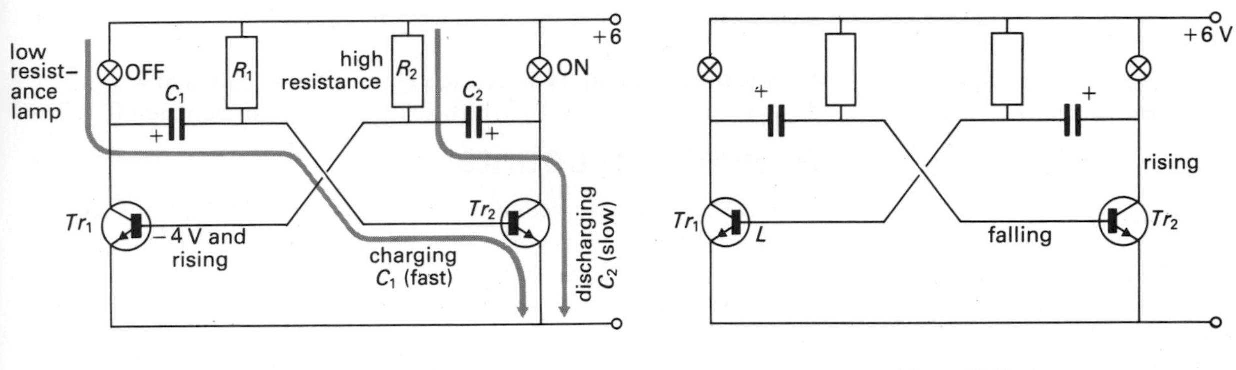

Figure 16.11 The astable multivibrator

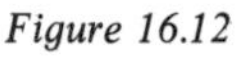

Figure 16.12

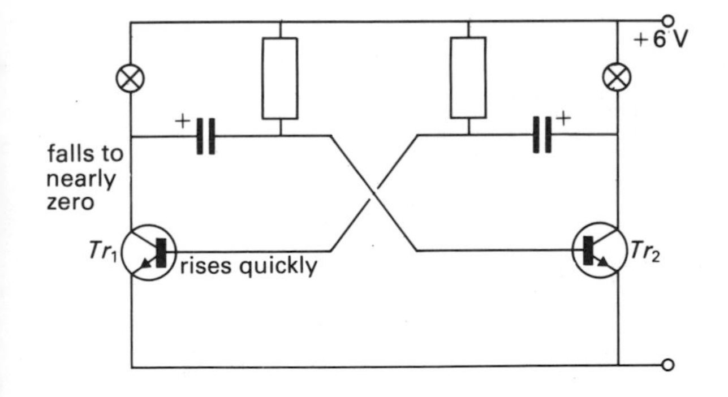

Figure 16.13

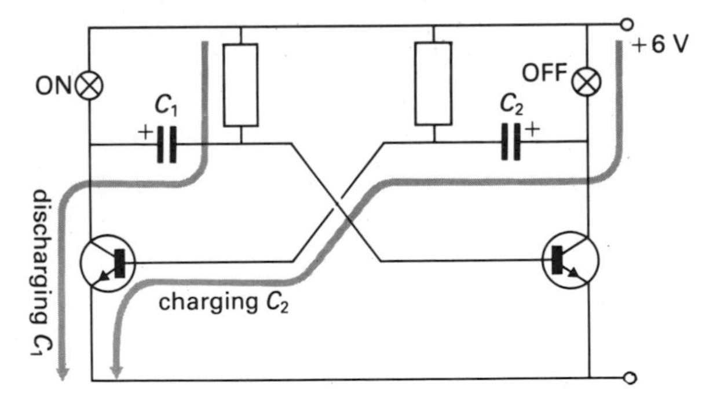

Figure 16.14

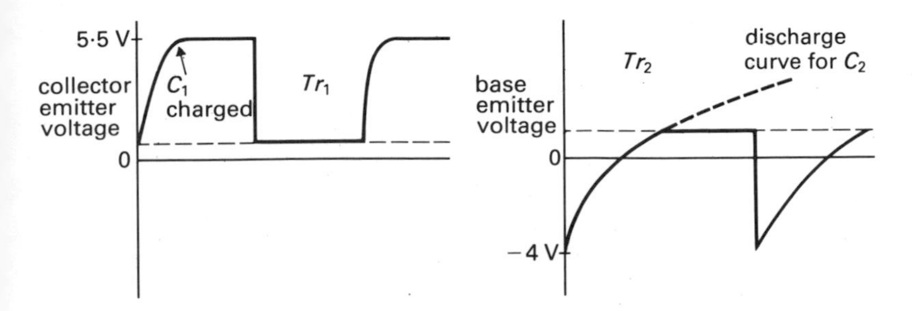

Figure 16.15 Oscilloscope waveforms obtained at the collector and base

Binary counters

Returning now to the bistable, notice that it takes two pulses of the square wave to change the state of each transistor. This means that one pulse will come out of Tr_2 collector for every two put in. The bistable does, therefore, divide by two; this is called binary division. Successive division by two can be obtained by means of a chain of bistables. It is clearer to see what is happening if each bistable is represented as a box with five terminals, as in *figure 16.16*. The meaning of the letters on the terminals is as follows:

a C is the input for pulses to be counted.

b Q and $\bar{Q}$ are the two output stages, one always being the inverse of the other.

c S is the set input, by which the Q output can be made 1.

d R is the reset input, by which the Q output can be made 0.

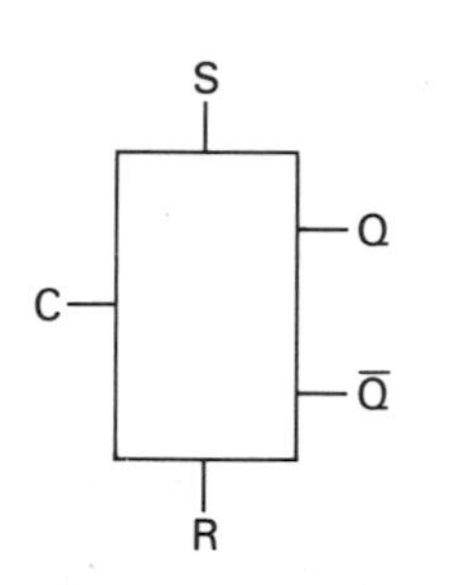

Figure 16.16 Symbol to represent a triggered bistable

A binary up counter

A three stage binary up counter is shown in *figure 16.17*. If the counter starts at 0000 (all lamps out), it will progress 0001, 0010, 0011, ... up to 1110, 1111, and then reset itself to 0000. It has counted, in binary, from 0 to 15 (decimal) and could repeat this millions of times every second, if necessary.

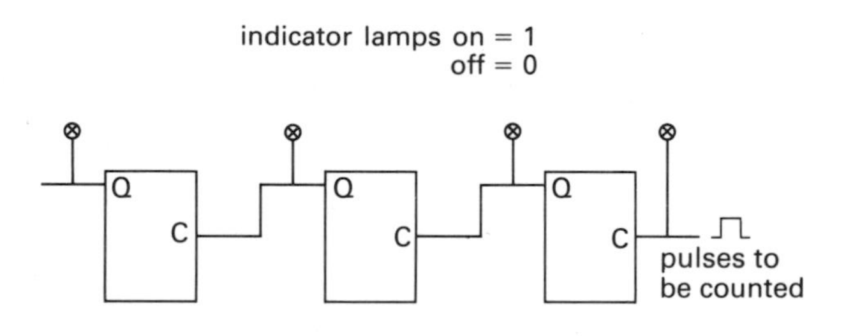

Figure 16.17 A binary up counter

A scale of ten counter

A binary counter can be made to count up to ten and then reset to zero by using the modification shown in *figure 16.18*. When the decimal

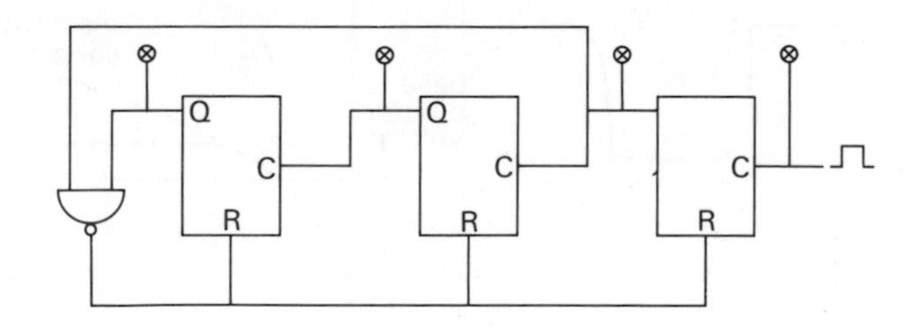

Figure 16.18

number ten is written in binary, it becomes 1010, so the appearance of a 1 at the first and third bistables can be used to reset all the bistables to 0. The next input pulse begins the count to ten again.

A binary down counter

If the $\bar{Q}$ output is used to trigger the following stage then the bistables will count down from any previously set number to zero. The circuit is shown in *figure 16.19*.

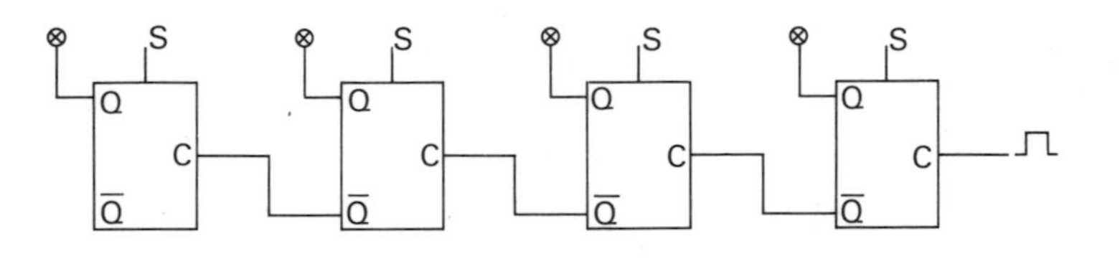

Figure 16.19

A shift register

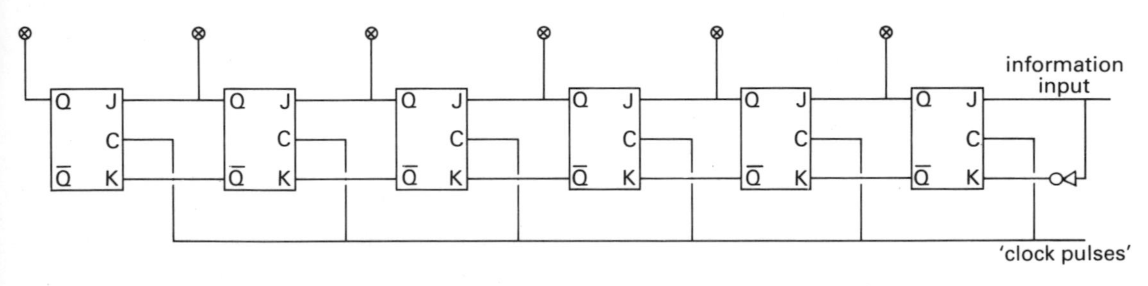

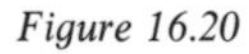
Figure 16.20

The chain of bistables shown in *figure 16.20* is made from *JK* bistables. These are different from those shown so far, because they have two inputs, *J* and *K*, in addition to the 'clock' pulse input. These bistables do not change their state until they receive a clock pulse, and then the result obtained depends upon the condition of the *J* and *K* inputs. The truth table for a *JK* bistable is shown in *figure 16.21*.

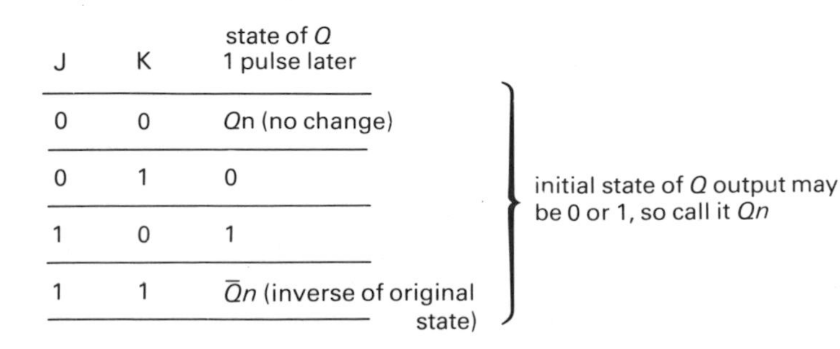

J	K	state of *Q* 1 pulse later
0	0	*Qn* (no change)
0	1	0
1	0	1
1	1	$\bar{Q}n$ (inverse of original state)

initial state of *Q* output may be 0 or 1, so call it *Qn*

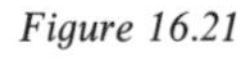
Figure 16.21

Initially the bistables are reset so that all the lamps are out. The use of a shift register is for the storage of information, so let the information be the 6 bit (**b**inary dig**it**) number 010110. It is fed in as follows:

0 * 1 * 0 * 1 * 1 * 0

where each * represents a single clock pulse. Subsequent clock pulses will shift the number to the left, thus:

0 1 0 1 1 0
1 0 1 1 0 0
0 1 1 0 0 0

By means of a circuit modification, it is also possible for the number to be shifted to the right.

The data latch

A data latch acts as a memory for the transfer of information from one part of a digital circuit to another. Instead of transferring constantly changing information, the latch acts as a 'buffer' and will only pass on its input information when it receives a pulse at its clock input. A 4 bit data latch is shown in *figure 16.22*.

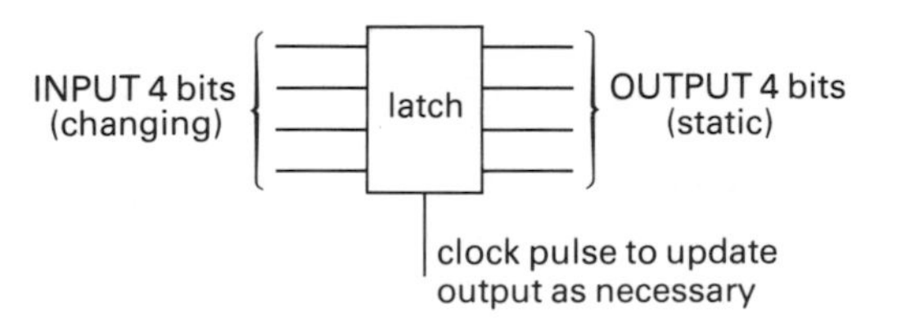

Figure 16.22 A 4 bit data latch

Numerical indicator tubes

The only type of indicator used so far to read the state of a bistable, is a lamp. As binary numbers are not as easy to read as decimal numbers, a counter is usually made to operate a decimal readout display. A typical seven segment display is shown in *figure 16.23*. In order to display, say, a number 2, segments *a b g e d* will be activated. The display itself may consist of seven filaments (now rare), 7 light emitting diodes or seven **liquid crystal** segments. A liquid crystal is a substance resembling a liquid in appearance, but when a p.d. is applied it takes on some degree of order in the molecular arrangement, like a crystal. When an electric field is applied to a liquid crystal, its light reflecting properties are also altered. Such a display requires very little electrical energy to operate it, but because it does not generate its own light, it must be illuminated by daylight or artificial light.

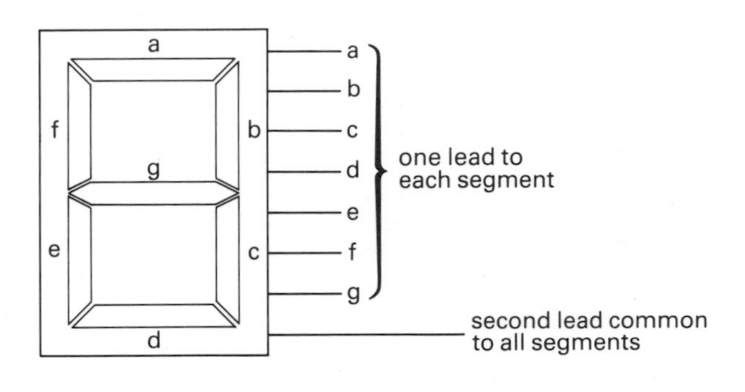

Figure 16.23 A seven segment display

Binary to seven-segment decoder

Although every decimal number has a binary equivalent, the method of converting one to the other is to represent every decimal digit by a 4 bit binary number.

Example:

binary number 110110
is just another way of writing.......... $2^5 + 2^4 + 0 + 2^2 + 2^1 + 0$
which is the same as .... $32 + 16 + 0 + 4 + 2 + 0$
giving a decimal number 54

The number 5 in binary form is 0101, and the number 4 in binary form is 0100, so in binary coded decimal (BCD) the number 54 would be:

0101 0100

One of the advantages of using BCD is the relative ease of storing 4 bit binary numbers, compared with the storage of a large bit binary equivalent.

In order to drive a seven segment display from a binary number, it is necessary to feed the information into a circuit which consists of a latch, a BCD to decimal decoder, and a seven segment driver to activate the required segments of the display.

A digital voltmeter

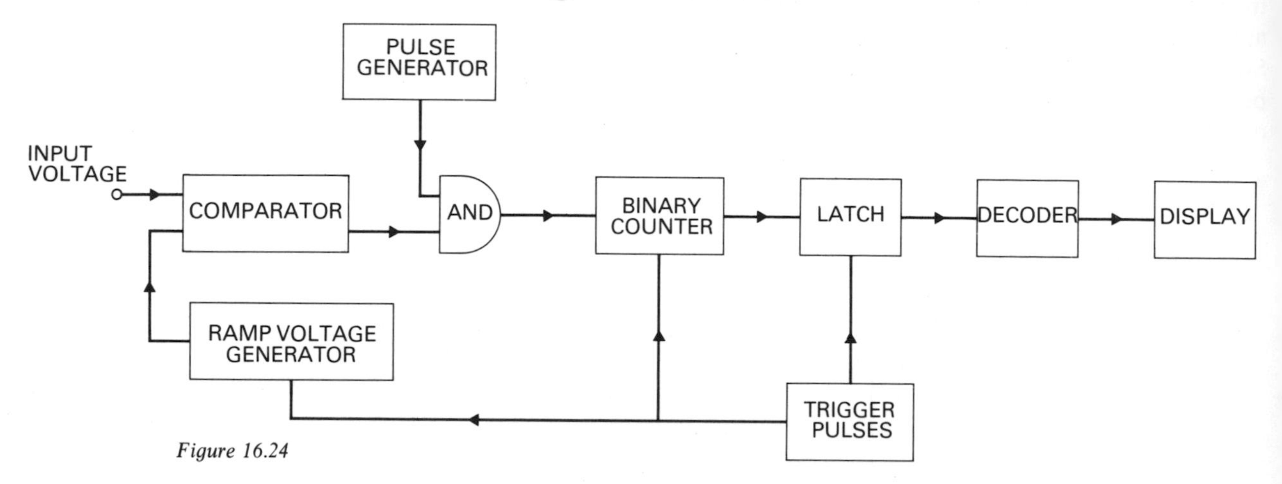

Figure 16.24

A number of the digital operations discussed so far can be seen in the simplified diagram of a digital voltmeter in *figure 16.24*. The order of events is as follows:

a a **trigger pulse** starts the **ramp** and **resets** the **counter** to zero.

b when the ramp voltage has risen as high as the input voltage, the **comparator** output changes from 1 to 0.

c the **AND gate** allows pulses to pass through it, until the comparator output becomes 0.

d the number of pulses allowed through the AND gate is proportional to the voltage being measured. This number is recorded in the counter.

e the **latch** passes this number to the **display**, where it is held until the next count is entered in the latch.

Oscillating circuits

Unit 17

Oscillations

We are very familiar with mechanical oscillations, such as the motion of a swing or pendulum. Water can also be made to swing to and fro by blowing briefly down one side of the U shaped tube as shown in *figure 17.1*. The time taken for the water to make one complete to and fro movement is called the **period**. The number of complete swings it makes in 1 second is called the **frequency**. The distance moved either side of its rest position is called the **amplitude** (*figure 17.2*).

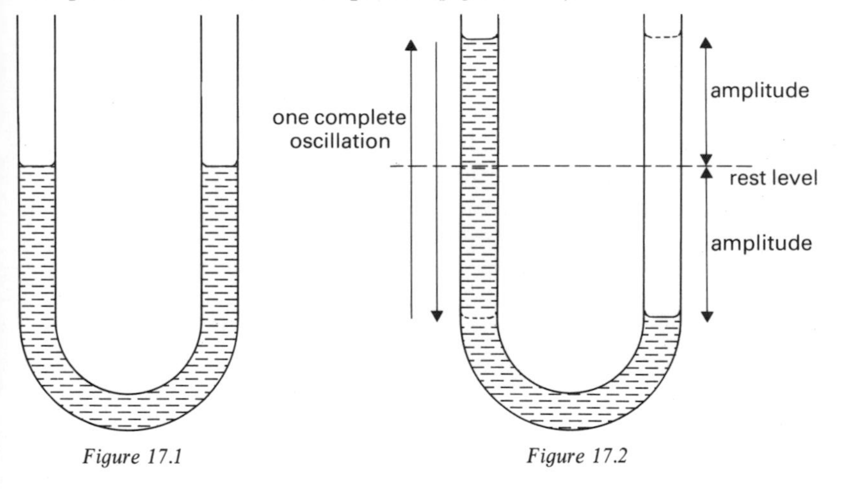

Figure 17.1 *Figure 17.2*

If the water is given a single push, the amplitude will gradually die away. This happens because the water loses energy as it moves, due to the resistance of the narrow tube. The resulting motion is called a **damped oscillation**, shown as a graph in *figure 17.3*. Note that the period (the time interval between one of the top peaks and the next) remains

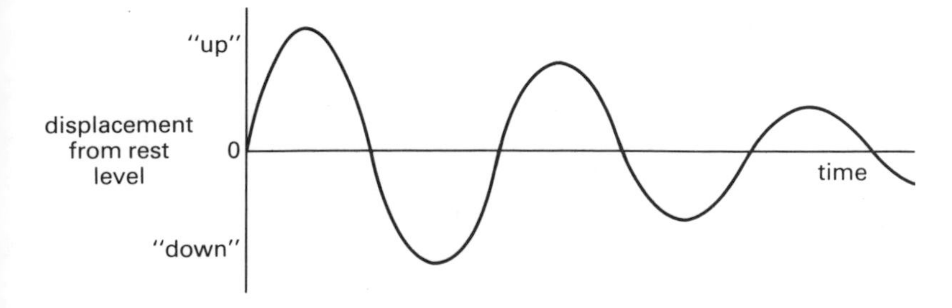

Figure 17.3 Graph to show damped oscillations

constant, even though the amplitude is decreasing. If the loss of energy is made up by pushing the water at exactly the right moments, the oscillation can be kept going at constant amplitude, as shown in *figure 17.4*. Only a small amount of energy needs to be supplied at each push, provided it is correctly timed. Most children have probably noticed that if they slide up and down in the bath at exactly the right frequency, the amplitude of the water can be quickly built up until it flows over the top of the bath! The right frequency for these pushes is the **same** frequency as the natural swing of the water. This effect is known as **resonance**.

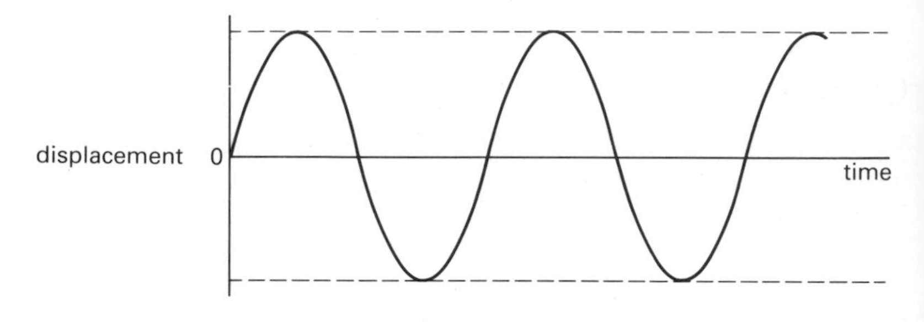

Figure 17.4 Undamped oscillations

Resonance

All examples of resonance can be summarised by the diagram shown in *figure 17.5*. Resonance will occur when the driving force has a frequency of f, or $\frac{1}{2}f$, or $\frac{1}{3}f$ or any simple fraction of f. The coupling provides the link along which energy is transferred to the driven system.

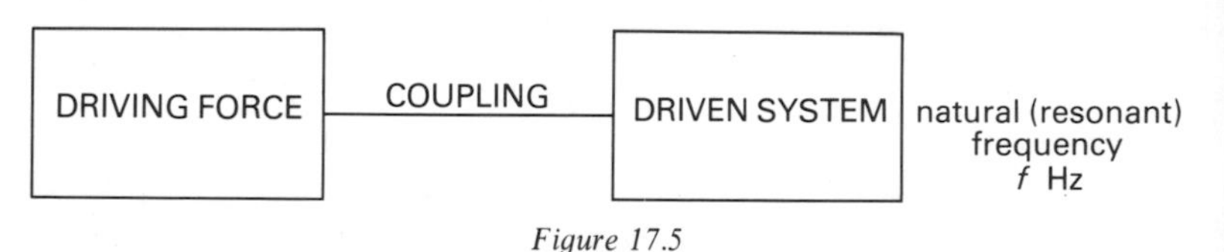

Figure 17.5

Electrical oscillation

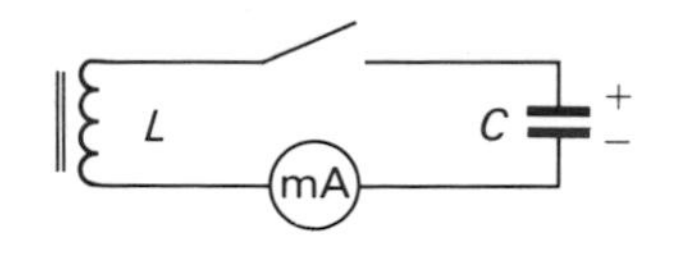

Figure 17.6 Circuit to demonstrate damped oscillations

In electrical oscillations, electrons swing to and fro in the conductors. *Figure 17.6* shows one way of producing damped electrical oscillations. The capacitor is first charged and then the switch is closed. The electrons swing round the circuit from one plate to the other, losing energy in the form of heat in the coil as they go. If L and C have large enough values, the frequency of the oscillations will be low enough to be followed by a milliammeter placed in the circuit. The meter's movement could be drawn as a graph, as shown in *figure 17.7*. The frequency f of the oscillation is given by the equation:

$$f = \frac{1}{2\pi\sqrt{LC}}\ \text{Hz} \qquad \begin{array}{l} L = \text{inductance in henrys} \\ C = \text{capacitance in farads} \end{array}$$

This frequency is called the **resonant frequency** of the circuit. At this frequency, the impedance of the circuit has its lowest value, but is not zero.

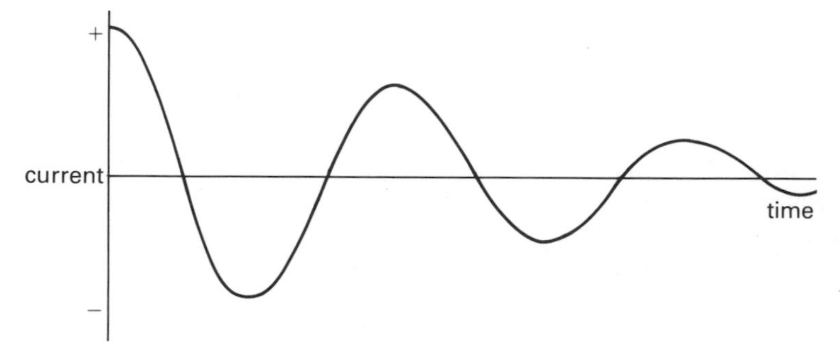

Figure 17.7 Current changes in a damped oscillation

Series and parallel resonance

There are two ways in which a capacitor and an inductor may be connected to an a.c. supply, as shown in *figure 17.8*. The following table compares the properties of these two circuits when they are at their resonant frequencies:

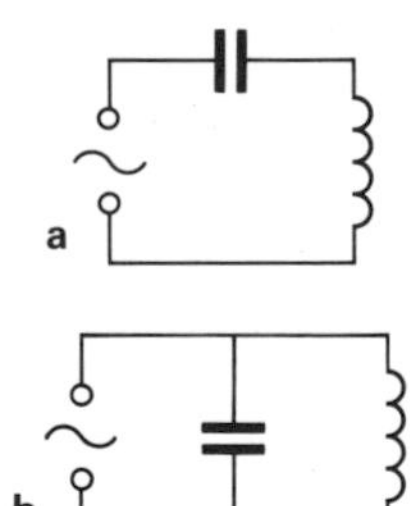

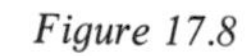
Figure 17.8

series resonance	**parallel resonance**
impedance is at minimum	impedance is at maximum
hence current is a maximum	hence current is a minimum
∴ called an acceptor circuit	∴ called a rejector circuit
resonant frequency $= \frac{1}{2\pi\sqrt{LC}}$	resonant frequency $= \frac{1}{2\pi\sqrt{LC}}$

Measuring resonant frequency

In order to keep the oscillations in *figure 17.6* going, a little energy must be fed in. This can be done with the aid of a signal generator, which itself produces electrical oscillations over a wide range of frequencies. If its frequency range covers the audible frequencies, it is called an audio frequency (AF) signal generator. The usual range of frequencies would be 10 Hz to 100 kHz. A radio frequency (RF) signal generator would cover the range of 100 kHz to 300 MHz, or thereabouts. *Figure 17.9* shows one way of connecting an AF signal generator, using its low impedance or 'loudspeaker' output terminals. The meter must be

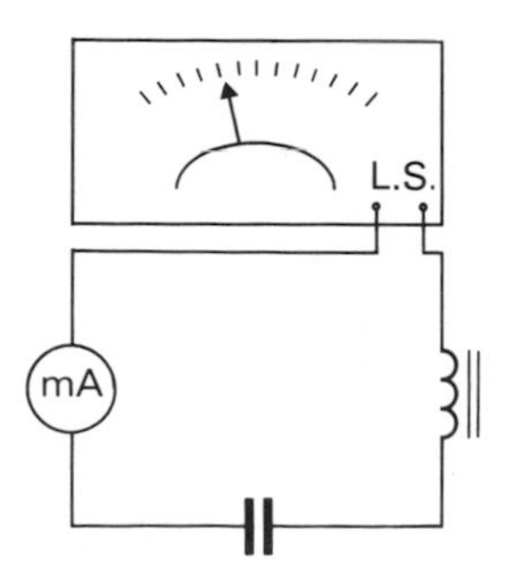

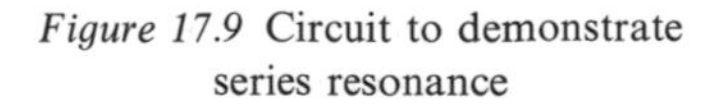
Figure 17.9 Circuit to demonstrate series resonance

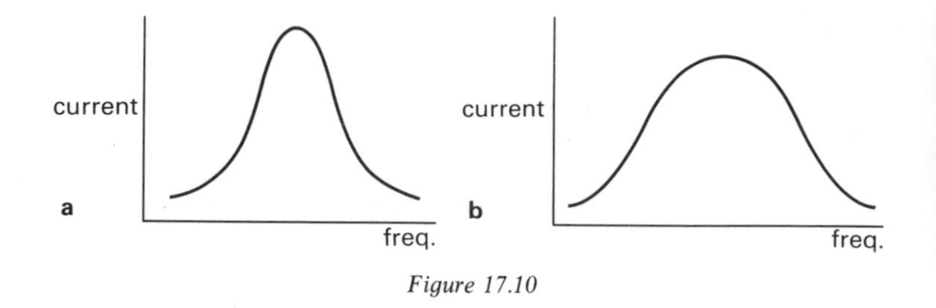

Figure 17.10

capable of reading alternating currents. If a graph is plotted of the meter reading against the frequency, it will look like *figure 17.10*. The lower the resistance of the circuit, the sharper will the curve fall away either side of the peak. Both sharp (*figure 17.10a*) and flat (*figure 17.10b*) resonant circuits have their uses.

Maintaining electrical oscillation

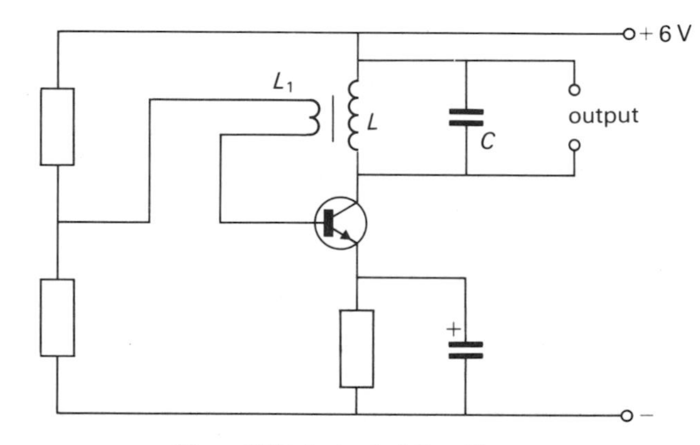

Figure 17.11 A simple *LC* oscillator

To keep the amplitude of electrical oscillation constant in a resonant circuit, a small amount of energy at the same frequency as the resonant frequency must be fed in via a suitable coupling. Compare the circuit shown in *figure 17.11* with the single stage amplifier described in unit thirteen, page 92, and notice how it meets all the requirements of:

a an amplifier.
b a tuned circuit of resonant frequency f.
c a source of energy of frequency f
d a suitable coupling
e positive feedback

Can you see what the effect would be of reversing the connections to L_1, or of increasing the value of C?

If fixed frequency RF oscillations are required, a **crystal oscillator** can be used. One possible circuit is shown in *figure 17.12*. Note that, at radio frequencies, the emitter bypass capacitor has a much lower value than in the AF oscillator circuit in *figure 17.11*. The crystal acts as a tuned circuit which feeds a portion of the output back to the input, and the output and the input are in step with each other.

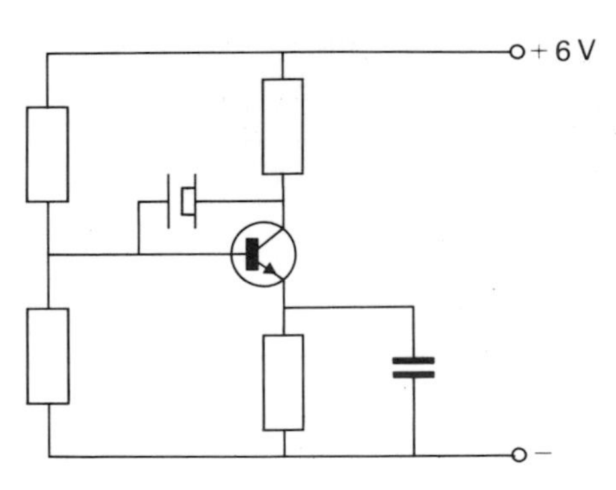

Figure 17.12
A simple quartz crystal oscillator

RC oscillators

The oscillator circuits shown so far are most suitable for generating high frequencies. If audio frequency oscillations are required, there are a number of suitable circuits which use resistors and capacitors only for the feedback and the tuning. They are called **RC oscillators** because they have resistor and capacitor coupling. They are so called **phase shift** oscillators, because the RC network alters the relative position of an a.c. wave. When two waves are out of step they are said to be **out of phase**. Two examples of phase shift oscillators are now given.

The Wien bridge oscillator

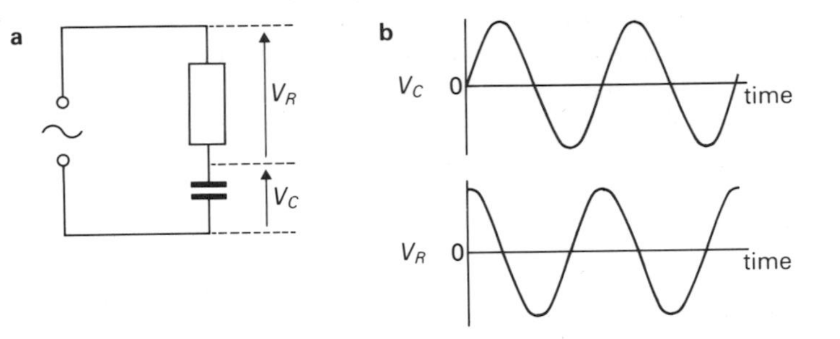

Figure 17.13 Phase shift in an *RC* circuit

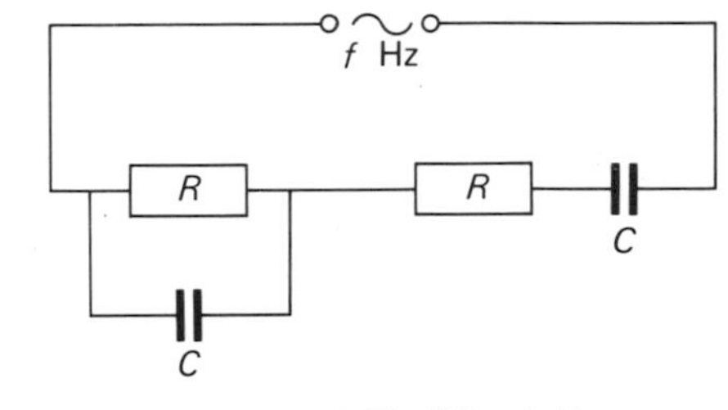

Figure 17.14 The Wien bridge

If an a.c. signal is applied to a capacitor and resistor in series, the voltage across the capacitor and across the resistor will be out of phase by a quarter of a wavelength, as shown in *figure 17.13*. The Wien bridge consists of two equal value resistors and two equal value capacitors connected as shown in *figure 17.14*. This circuit produces a phase shift at all frequencies except when $f = 1/2\pi CR$. If this circuit is used to provide positive feedback to an amplifier, it will do so at a frequency of $1/2\pi CR$ Hz, because at all other frequencies the signal will be fed back out of step and can, therefore, never provide the reinforcement to maintain the oscillation. The circuit shown in *figure 17.15* shows how an operational amplifier can be used as a Wien bridge oscillator. The frequency can be changed by altering the values of both sets of resistors labelled *R*.

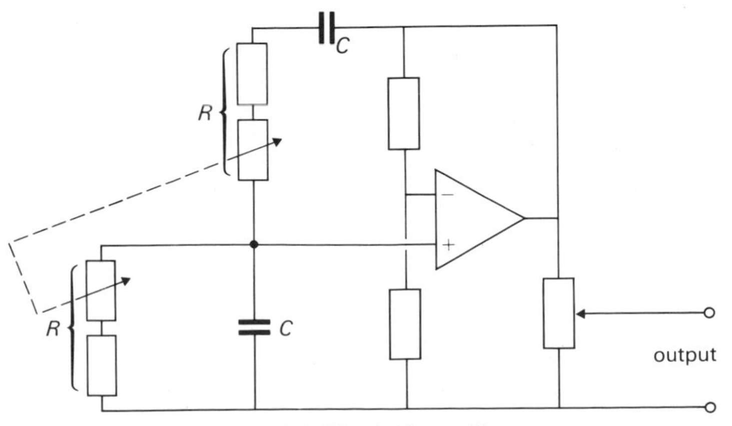

Figure 17.15 A Wien bridge oscillator

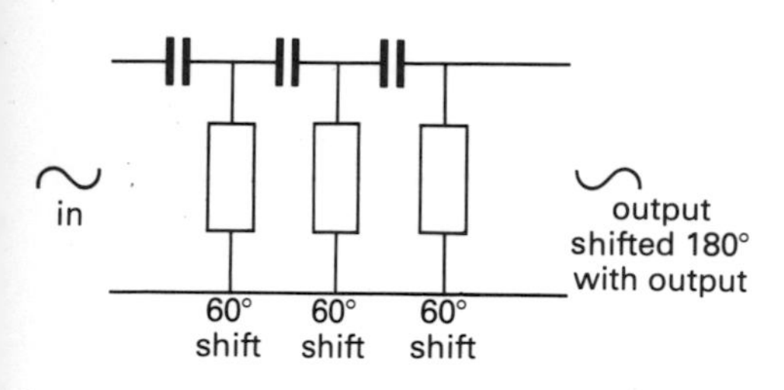

Figure 17.16

The phase shift oscillator

The network of three capacitors and three resistors shown in *figure 17.16* will produce a phase shift of exactly 180° at one frequency only. That frequency f_r is given by the equation

$$f_r = \frac{0{\cdot}065}{CR}$$

If the amplifier also introduces a phase shift of 180° then the output will be in step with the input, because a shift of 360° between two waves brings them back into step again. As the output is now in step with the input, it will reinforce it. However, the RC network introduces a loss of about thirty, so the amplifier must have a gain of at least thirty in order for oscillations to be maintained at a frequency of f_r. This circuit is only suitable for operation at a single frequency, because to vary the frequency, either the three resistors or the three capacitors must be varied simultaneously.

Relaxation oscillators

All of the oscillators described so far are capable of producing a pure sine wave output. Relaxation oscillators produce non-sine wave outputs. The two most useful are **sawtooth** and **square wave** oscillators. A sawtooth voltage (*figure 17.17*) is used in an oscilloscope as the time base (see unit six). A square wave (*figure 17.18*) is very useful for certain testing purposes, for example, if an audio amplifier is fed with a square wave input and it gives a square wave output, this means the amplifier is working well.

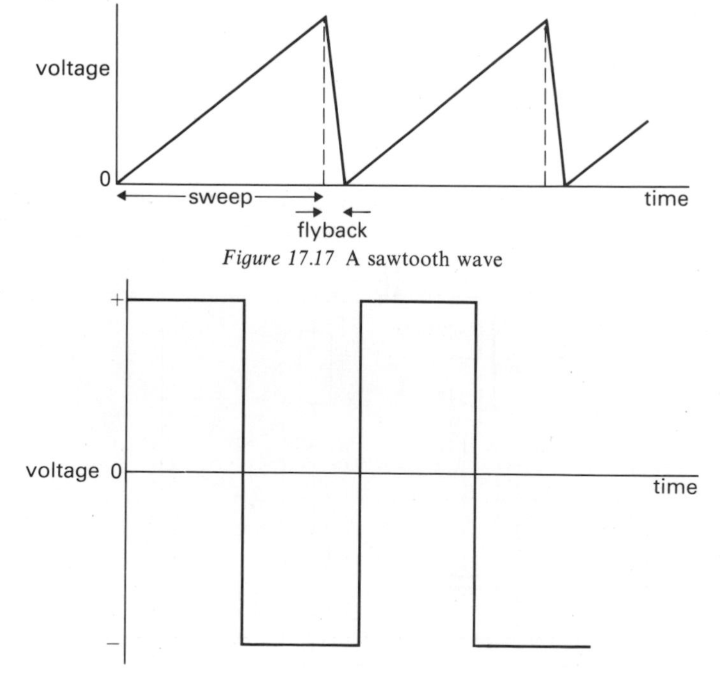

Figure 17.17 A sawtooth wave

Figure 17.18 A square wave

Generating square waves

Most AF sine wave generators will also produce square waves. This is done by taking the sine wave, amplifying it and clipping off the top and bottom of the peaks, as shown in *figure 17.19*. A multivibrator (see unit sixteen) also produces squares waves, and it is possible to alter the **mark-space ratio** as shown in *figure 17.20*.

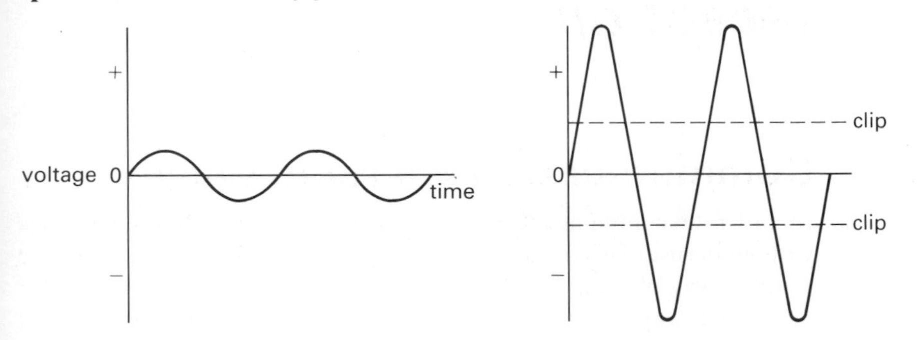

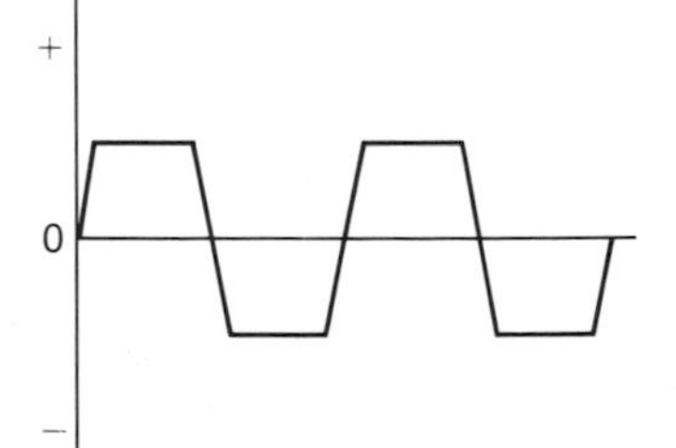

Figure 17.19 Making square waves from sine waves

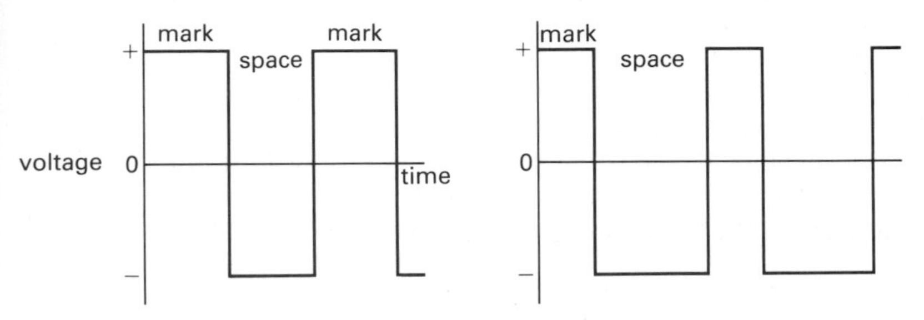

Figure 17.20 Varying the mark-space ratio

Pulse generators

A pulse generator can be considered as an oscillator which produces a short burst of energy, followed by a relatively longer gap before the next burst. A graph of the output would look something like *figure 17.21*. Radar works by picking up the echo of short pulses of very high frequency waves, sent out by a transmitter and reflected by distant objects such as aircraft or ships.

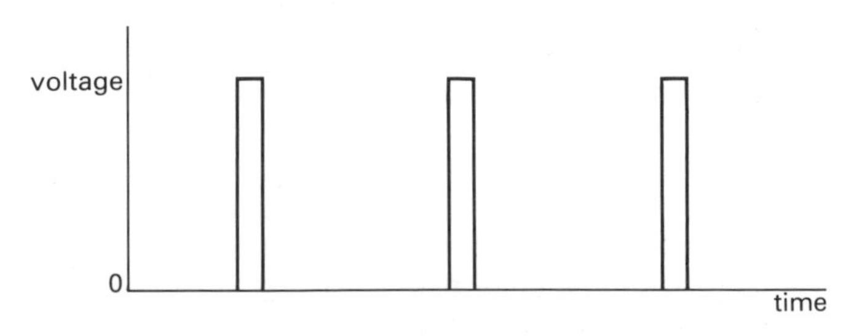

Figure 17.21 Pulse waveforms

Unit 18

Radio waves and their reception

Communication via radio waves

In order to communicate over distances and through solid objects without the use of wires, radio ('wireless') waves are used as a **carrier**. The simplest way of using these waves is to turn them on and off in some prescribed way, such as in the Morse code. This is the form of transmission known as the **interrupted carrier** wave, and it can sometimes be heard on short wave radio bands as a 'hissing' Morse code. A specially modified receiver is necessary to turn the hiss into the more familiar musical note of the Morse code.

In order to transmit speech or music on a radio wave, the carrier must be **modulated**, that is, it must be made to change in some way corresponding to the varying frequencies of speech or music. The two types of modulation used for commercial broadcasts are **frequency modulation** (FM) and **amplitude modulation** (AM).

Frequency modulation

Frequency modulation is used by the BBC to transmit all their VHF radio programmes, and for the sound of UHF television. For radio, it consists of a carrier wave of constant amplitude with a frequency somewhere in the range 88–95 MHz, according to the station and the transmitter location. This carrier frequency is then modulated by swinging its frequency up to 75 kHz above and below the carrier frequency. The rate at which it is swung is determined by the frequencies of the speech or music. The range of the swing is determined by their loudness, the full 75 kHz swing corresponding to maximum loudness. The method of receiving FM transmissions is beyond the scope of this book but it is worth noting its chief advantages:

a FM broadcasts can carry higher audio frequencies than a.m. and so give a more faithful reproduction of the original sound.

b the VHF radio band used for FM is less congested, so interference from other stations of neighbouring frequency is unlikely.

c most other types of interference, such as that from car ignitions and electric motors, are amplitude modulated and should not be received by a well adjusted FM radio receiver.

Amplitude modulation

Amplitude modulation is used for commercial broadcasts on long, medium and short wave bands. It consists of a carrier wave of fixed frequency (e.g. 200 kHz for BBC Radio 4), which is modulated by varying its strength (**amplitude**). The rate at which the carrier strength varies is determined by the frequencies of the speech or music. The amount the carrier strength varies is determined by their loudness. Thus the maximum permitted loudness corresponds to the carrier amplitude ranging between zero and twice its unmodulated amplitude. Silence corresponds to constant carrier amplitude (*figure 18.1*). The modulated carrier wave is still composed entirely of radio frequencies and can therefore be radiated from a radio transmitter. The audio frequencies have imprinted themselves by altering the shape of the **envelope** of the RF carrier.

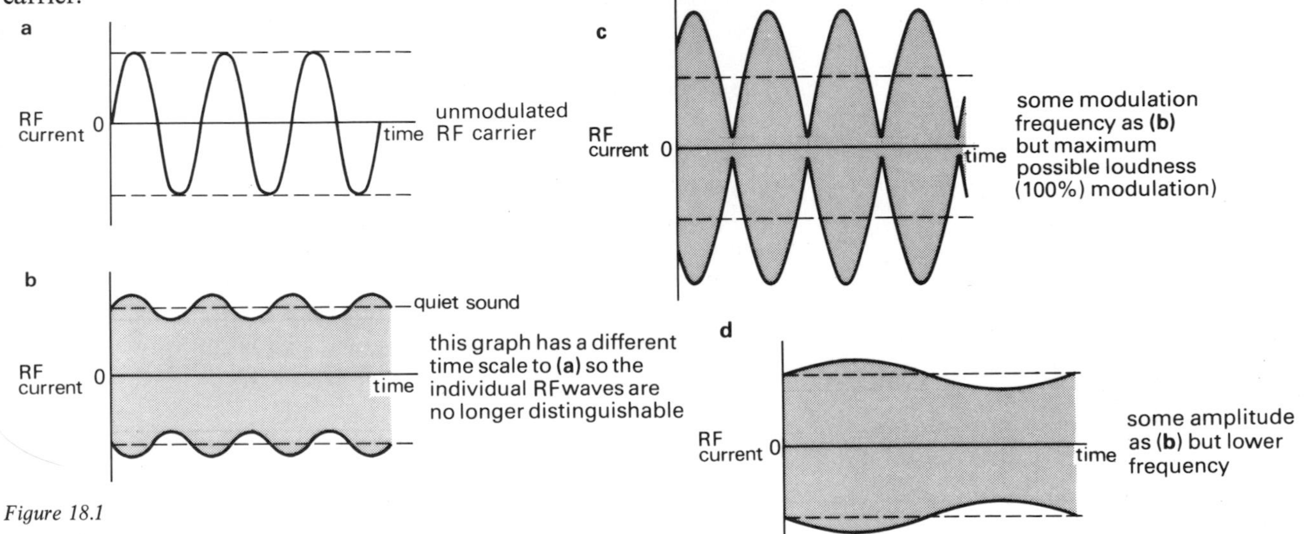

Figure 18.1

Reception of AM waves

Reception of AM waves involves the following steps:

a reception by the aerial.
b tuning to the required carrier frequency.
c amplification of the modulated carrier.
d demodulation (detection).
e amplification of the demodulated signal and replay through loudspeaker.

These steps are now described in a little more detail:

The aerial

Any length of wire will have tiny e.m.fs. induced in it owing to the presence of radio waves. For the e.m.f. to be a maximum a properly designed aerial or antenna is required. The most common types of aerial are the **dipole**, the **vertical whip** and the **ferrite rod**.

Right: Folded dipole aerial

wires to receiver

Figure 18.2
A simple dipole aerial

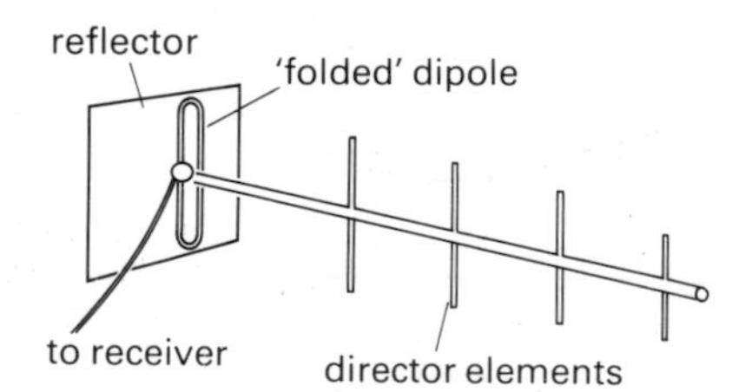

Figure 18.3
Folded dipole with director and reflector elements

The dipole may take the simple form as shown in *figure 18.2* or be combined with directing and reflecting elements as shown in *figure 18.3*. The whip aerial can be seen on cars, and the ferrite rod aerial is used in most portable radio receivers.

In the case of the dipole and whip aerials, the maximum output will be obtained if the aerial is of the correct length and is correctly aligned on the transmitter. For a dipole, the correct length is one half the wavelength of the carrier wave whose reception is required. To calculate the wavelength of these waves the following equation is used:

$$V = f\lambda$$

where V = velocity of the radio waves (300 000 000 m/s)

f = frequency in Hz

λ = wavelength in metres (λ is pronounced lambda)

Example Consider the reception of VHF radio signals at about 90 MHz.

$$V = f\lambda$$

$$300\,000\,000 = 90\,000\,000\,\lambda$$

$$\therefore \quad \lambda = \frac{300}{90} = 3{\cdot}33\text{ m}$$

This means the dipole length must be about 1·66 m overall.

To obtain the correct alignment it is necessary to know the position of the transmitter and whether its aerial is vertical or horizontal. For the London area, the transmitter is at Wrotham Heath and the waves are horizontally polarised from a horizontal aerial. As a further example, try to calculate the dipole length for a UHF television receiver operating at about 600 MHz.

To understand the ferrite rod aerial, we must look at the next stage of the reception process, namely the tuning.

Tuning the receiver

Even though the aerial may be accurately adjusted in length to a required frequency it will pick up many other frequencies. In order to pick out just one of these frequencies we make use of resonance. The aerial voltage (the driving force) is coupled to a tuned circuit (the driven system, see *figure 17.5* in unit seventeen). In its simplest form, the circuit is shown in *figure 18.4*. The aerial is **directly coupled** to the tuned circuit, the resonant frequency being adjusted by the tuning capacitor C. A circuit as simple as this will not completely separate carrier waves which are close in frequency; we say this circuit is not very **selective.**

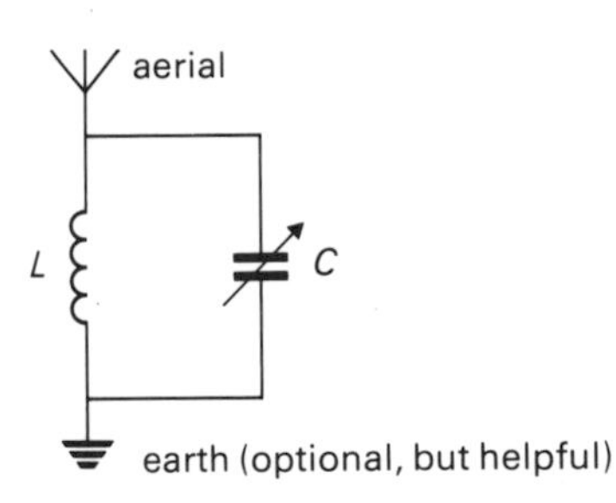

Figure 18.4 A simple tuned circuit

The ferrite rod aerial is commonly used for delivering a signal to the tuned circuit without the need for an external aerial. **Ferrites** are a group of non-conducting magnetic materials which, because of their special magnetic properties, tend to concentrate the radio waves passing nearby. The ferrite rod and the coil, together form the inductance, L, which is tuned by capacitor C, as before (*figure 18.5*). The size of the rod is not critical and the larger its volume, the greater the effect it produces. It is a **directional** aerial and must be lined up with the transmitter.

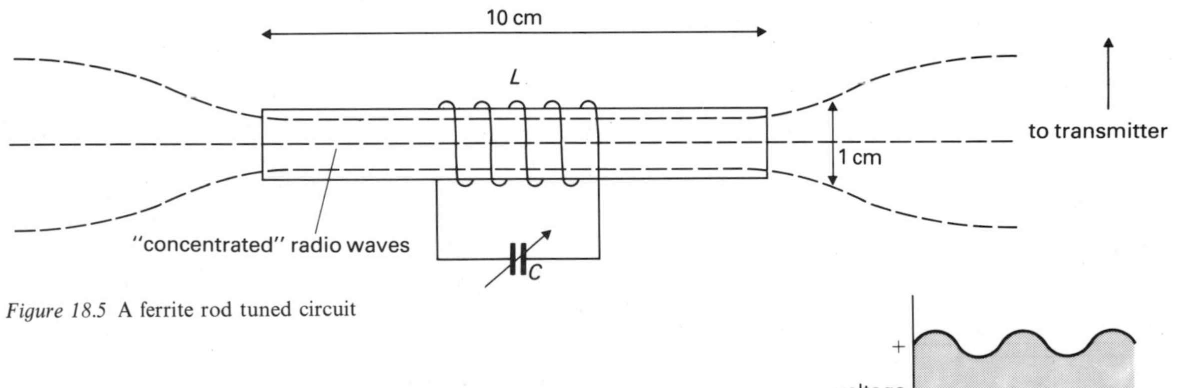

Figure 18.5 A ferrite rod tuned circuit

Radio frequency amplification

All commercial radio receivers have stages of amplification after the carrier has been tuned, but the details of this are beyond the scope of this book.

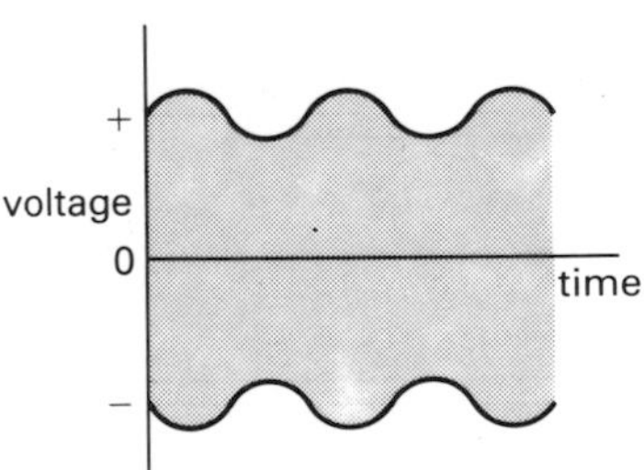

Figure 18.6 Modulated r.f. carrier as received by the tuned circuit

Detection

In order to reproduce an AM or FM radio broadcast, it is necessary to extract the audio frequencies which modulated the carrier in the first place. This process is called **demodulation** or **detection**. The first stage in the detection of AM, is to remove one half of the modulated carrier (*figures 18.6–7*). All that is required now is to filter out the remaining RF carrier and to restore the average d.c. level to zero. The circuit shown in

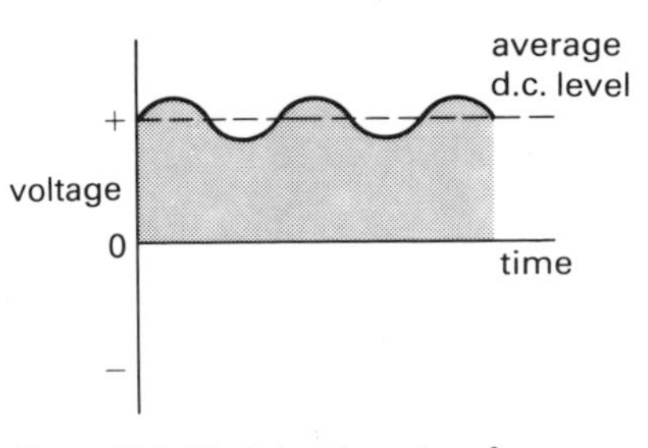

Figure 18.7 Modulated carrier after detection

figure 18.8 provides both rectification and RF filtering. The AF output from this should correspond in frequency and relative amplitude to the original modulating signal.

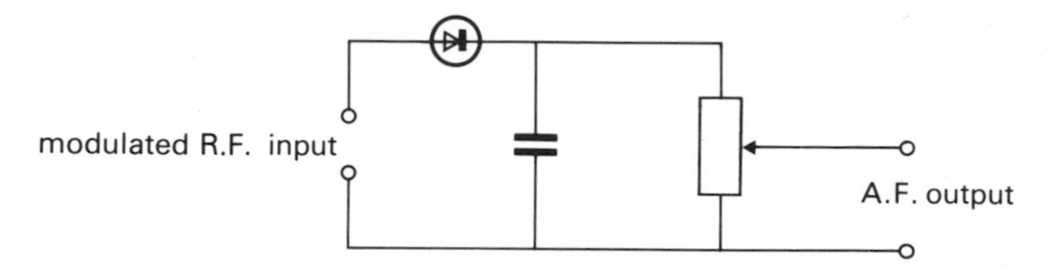

Figure 18.8 A diode detector circuit

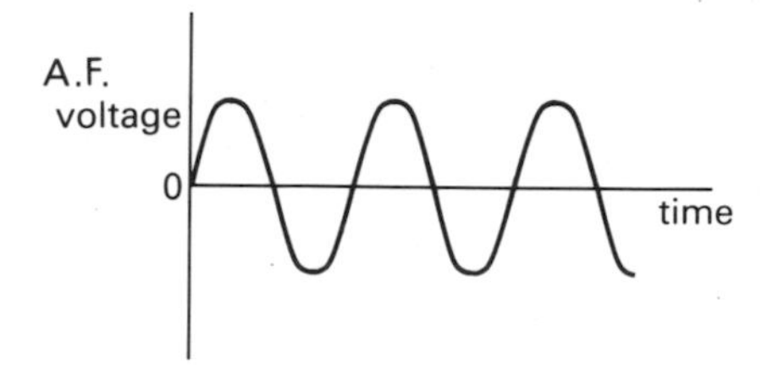

Figure 18.9 Voltage output from the detector circuit

Amplification and replay

Once the audio signal has been obtained, amplification can follow in the usual way to drive a loudspeaker. A suitable pair of high impedance headphones (**not** the type used for listening to hi-fi) may be connected as the detector load. We now have all the units for a diode receiver, the simplest way of receiving AM broadcasts.

The diode receiver

The complete circuit for a diode receiver is shown in *figure 18.10*. It will work well on any strong medium wave station. As it does not have any amplification at RF or AF, the loudness of the signal in the headphones will depend upon the length of the aerial, and the power and proximity of the transmitter. A simple receiver such as this is not very selective and you are quite likely to hear two or more stations at the same time.

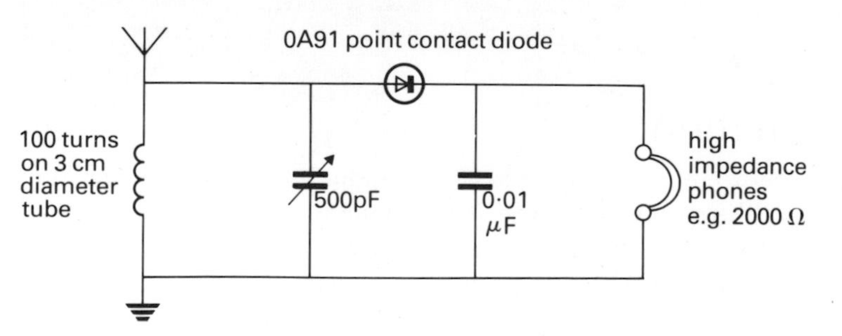

Figure 18.10 A complete diode receiver

Tape recording

Unit 19

Monophonic tape recording

A monophonic recording carries only one **channel** of information. At the studio there may well be more than one microphone in use, but the outputs of all the microphones are fed into a **mixer**, where they can have their levels adjusted and mixed to give a single output channel, with the required **balance** between the various sources of sound.

Recording studios usually make a **master tape** recording first. This can then either have tape copies made from it, or be used to prepare a disc recording (see unit twenty). The greatest advantages of tape recording are the possibility of re-using tapes and the possibility of editing a recording after it has been made, by cutting the tape and **splicing** it together again with a special sticky tape. This can be used to remove unwanted material or to change the order of the original recording.

The first magnetic recorder was made in about 1900, before the existence of electric amplifiers. The sound was recorded on steel wire instead of tape. German radio stations started to use magnetic tape recorders in 1938, using paper tape coated with particles of iron. In 1947, the 3M company in America introduced the sort of tape which we know today – a thin, tough, flexible plastic base covered with finely powdered iron oxide (Fe_3O_4) or chromium oxide.

The recording machine

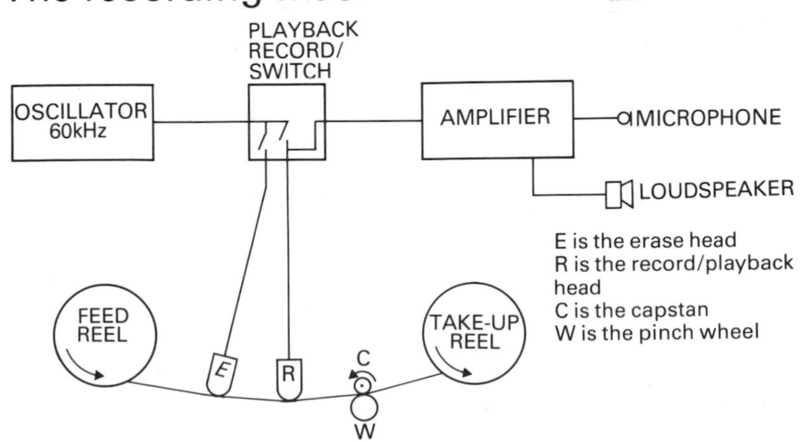

Figure 19.1 The main parts of a tape recorder

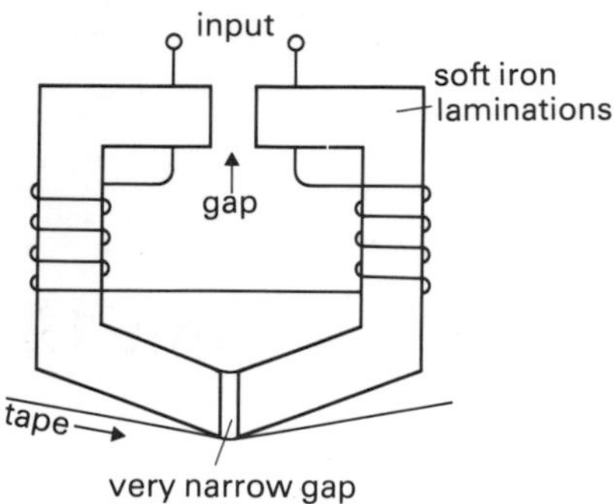

Figure 19.2 A tape recording head

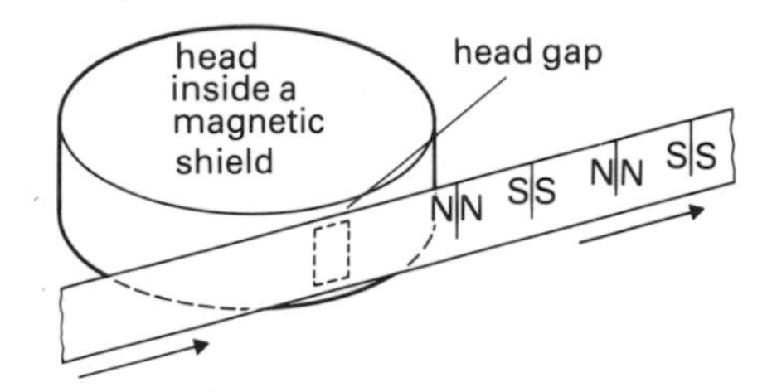

Figure 19.3 Producing the magnets on a tape during recording

The essential components of any tape recorder are illustrated in the block diagram shown in *figure 19.1*. The record–playback and erase heads are really just electromagnets, with a construction similar to that shown in *figure 19.2*. On top quality tape recorders, the gap in the record–playback head can be as narrow as 1 micron (one millionth of a metre). On these machines it is quite likely that a separate head will be used for record and playback, which gives the useful ability of a machine to replay the recording about half a second later. This enables the recording engineer to **monitor** the recording that has actually gone on the tape.

While the machine is recording, the varying e.m.f. from the microphone is amplified in the tape recorder and fed to the coil of the record head. Here it drives a varying current and so produces a varying magnetic flux in the tiny gap. As the tape passes this gap, a series of permanent magnets is produced on the tape as shown in *figure 19.3*. The spacing of these magnets will depend upon the frequency of the current changes in the recording head. The strength of these magnets will depend upon the strength of the recording current, which in turn depends upon the frequency and strength of the original sound reaching the microphone. As there is a limit to the strength of magnetism which can exist in any magnetic material, a **recording level** or VU (*V*olume *U*nits) meter must be used to indicate when the limit has been reached. Recordings made above this level will be **distorted**. *Figure 19.4* illustrates the effect of different frequency recording currents.

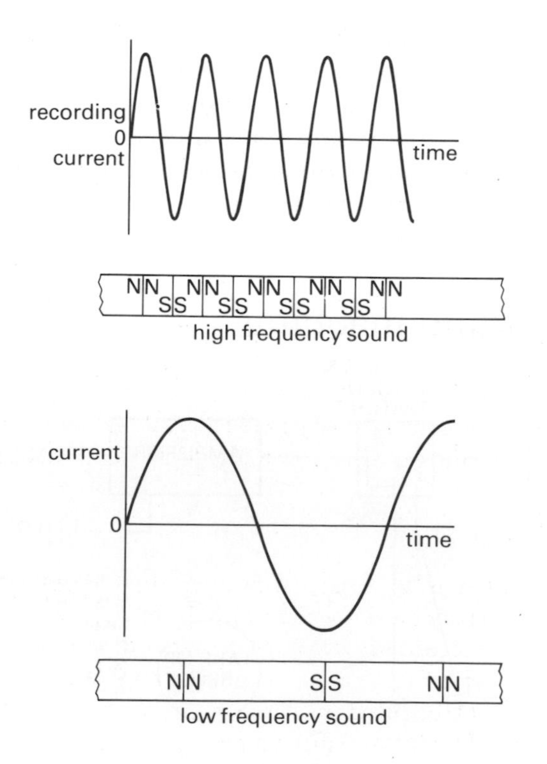

Figure 19.4
High and low frequency signals on a tape

Tape speed and frequency response

If the tape speed is low, e.g. 9·5 cm/s, the high frequency sounds will produce magnets on the tape which are very short. These magnets tend to lose their magnetism immediately, so making it difficult to record frequencies above about 16 kHz. If the tape is run past the head at 19 cm/s, then frequencies up to 24 kHz can be recorded. There is less of a problem with low frequencies, where the limitation is more likely to be the loudspeaker used for the replay.

Erasing the tape

When the machine is recording, the erase head is supplied with a high frequency alternating current (about 60 kHz). The construction of the erase head is similar to the record/playback head, but the front gap is larger. The current supplied to it is also greater. Its function is simply to remove any magnetism from the tape before it passes over the record head. The best method of removing magnetism from an object is to pass it through a coil in which an alternating current is flowing. Accidental erasure of a tape is not likely to happen because the erase head only works while a recording is being made. However, a tape can be spoiled, if not actually erased, by bringing it close to any sort of magnet.

The need for recording bias

If the recording head is supplied with the recording current only, the result on replay will be distorted. This is the inevitable result of the magnetic properties of the tape. The method used to overcome this distortion is to feed into the recording head, at the same time as the signal to be recorded, an alternating current with a frequency of about 60 kHz, usually taken off the erase current. This is called the **recording bias**. The amount of bias required is quite critical; if it is too low the recording will be distorted, and if it is too high there will be a loss of high frequencies from the recording. As the correct amount of bias will depend upon whether iron oxide or chromium oxide tape is being used, many tape recorders have a switch to allow the correct bias to be selected for each tape. The action of the recording bias may be considered as that of 'shaking up' the molecular magnets in the tape, so that they can respond more uniformly to the changes in the recording head magnetism.

Moving the tape

If the tape recorder is to replay accurately the sounds fed into the microphone, the replay must be made with the tape moving at the same speed as that of the recording. It is also necessary for the speed to remain constant over a period of time, to within 0.5% or less. Any variation greater than this may be heard as a variation of pitch, especially on long steady notes played by musical instruments. This variation has come to be called **wow**, if the frequency wobble is slow, and **flutter**, if the wobble is fast.

The speeds at which tape is pulled past the heads has been fixed by international agreement. On domestic tape recorders the speeds are usually 19 cm/s and 9·5 cm/s for open reel machines, and 4·75 cm/s for cassette machines. The tape is pulled along by being tightly sandwiched between the rotating **capstan** and the **pinch wheel** (*figure 19.1*). The capstan carries a heavy flywheel, turned by a constant speed motor. The take-up reel is also driven, often by the same motor. More expensive machines may have three separate motors, one for the capstan and one for each reel.

The tape is kept in close contact with the record/playback head, often by means of pressure pads. If the head becomes clogged with oxide, rubbed off from the tape, this close contact is lost and the quality of the recording suffers. Regular cleaning of the heads with cotton wool soaked in alcohol will prevent this. For cassette tape recorders, there is a special head cleaning cassette which can be run through the machine from time to time, to remove the oxide from the heads.

Replaying the recording

During replay the tape is pulled past the head, which may be a separate head to the right of the record head, or, more often, the same head as was used for the recording. As each magnet on the tape passes the gap, it generates a tiny voltage in the coils. This voltage will vary at the same frequency as the current which produced it during the recording. The tiny voltage is now amplified and fed to a loudspeaker, where the original sound should be faithfully reproduced.

Two track recording

Most monophonic tape recorders have **half track** heads, which means that only half the width of the tape is used. In order to use the other half, the cassette or reels are turned over and the tape run through again. Editing the tape by cutting and joining is no longer possible if two or more tracks have been recorded on the same tape.

Stereo recording

A stereo tape recorder requires special record–playback and erase heads which will make two recordings, at the same time, on different tracks of the tape. Each head has two separate electromagnets, and the gaps cover just under one quarter of the tape width, as shown in *figure 19.5*.

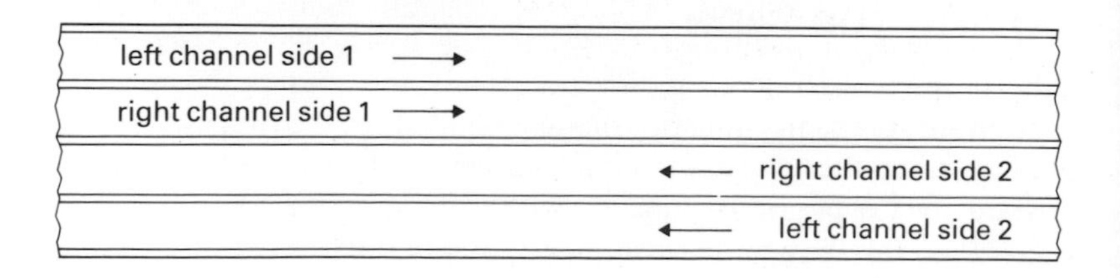

Figure 19.5 Arrangement of tracks on a 4 track stereo recording

Cassette recorders

When cassette recorders were first introduced, they were never intended to be anything other than compact, portable tape recorders for general purpose use. As a result of improvements in tape quality, head design and modern electronics, these machines are now being sold as part of serious 'hi-fi' systems. The best machines are capable of recording frequencies up to 17 kHz which, for a tape speed of only 4·75 cm/s, is quite remarkable.

Dolby noise reduction

A special feature of many good quality tape recorders is their use of the Dolby noise reducing system. This works by increasing the level of all high frequency signals during the recording, and reducing their level back to normal during playback. The overall effect of this is to reduce the tape 'hiss', which can be so objectionable during quiet moments of playback.

Video tape recorders

A specially designed tape recorder is now available on the domestic market for the recording and replay of television programmes. The television signal, after detection in the receiver, carries frequencies from 50 Hz to above 5 MHz. If a conventional tape recorder was to be used to record at 5 MHz, a very high tape speed would be required. This problem has been overcome by making the record head rotate and 'scan' the tape in diagonal strips as the tape moves forward (*figure 19.6*). Different manufacturers have evolved different systems for recording television, and so their tapes and equipment are not interchangeable. The ultimate aim of the manufacturers is to provide a machine which will have a long playing time on a relatively cheap disc or tape. The disc or tape should suit machines of any manufacture, just as they do for sound recordings.

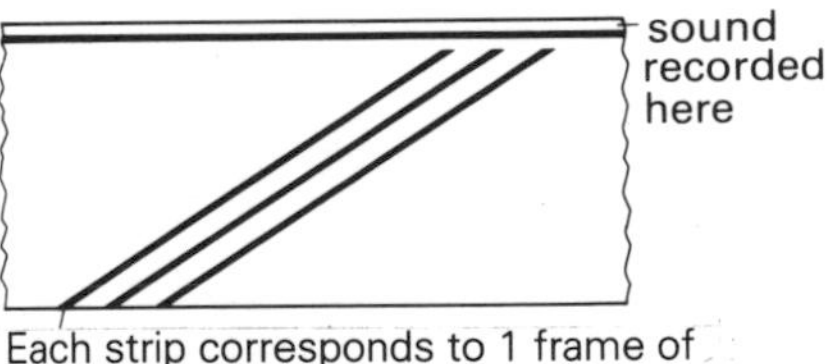

Figure 19.6
A video tape recording

Digital recording

Recording engineers are now generally agreed that the present methods of analogue recording have reached the limit of quality. An analogue signal is one whose amplitude varies continuously. Low amplitude signals tend to become lost in the background noise. If the analogue signal is coded into a stream of pulses of constant amplitude, these pulses can be recorded at maximum amplitude, thus eliminating the problem of background noise.

The first stage in the introduction of digital recordings is to replace the analogue studio tape recorder by a digital recorder. This will produce the master tape and will effect a worthwhile improvement in clarity, but if the tape is to be used to cut a disc, limitations to quality will still exist, owing to the natural surface roughness of disc plastic, and imperfections in the path of the needle along the groove. Editing of digital tapes is carried out by copying the required sections on to another tape, and not by cutting and joining.

Unit 20

Disc recording and replay

Cutting the disc

cutting head
mechanical drive

Figure 20.1
A disc cutting lathe

As was mentioned in unit nineteen the recording company cuts the disc recording from a master tape recording and not generally from a live studio performance, although when disc recording started, in the early part of this century, there were no tapes and the recordings were made direct. (It is interesting to note that in the quest for extreme high fidelity a few discs have been produced recently which have been cut directly, with no intermediate tape recording.) The first disc recordings were purely mechanical, the recording stylus being moved by the energy of the collected sound waves. Around 1925, electric amplifiers became available and the first electric recordings (at 78 rpm) were made. The 33 rpm microgroove record made its first appearance in 1950.

After the tape has been edited the preparation of the disc can begin. The first copy is made on an acetate disc, the groove being cut by a heated chisel of approximate width 0·05 mm. The chisel is driven towards the centre of the rotating acetate disc, and at the same time is moved a little from side to side by the audio frequency currents which it is recording (*figure 20.1*).

During quiet passages of the recording, the mechanical drive to the cutting head produces about 150 grooves to the centimetre, whereas on louder passages this has to be reduced to about 50, in order that they do not break into each other (*figure 20.2*).

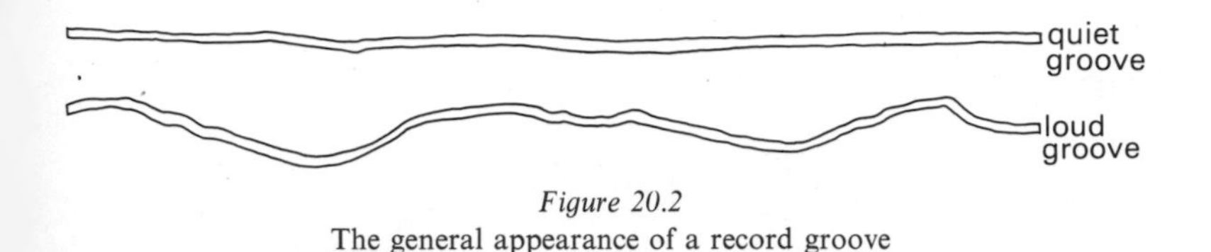

Figure 20.2
The general appearance of a record groove

After cutting, the master disc is electroplated with nickel making a 'negative' copy. This in turn is used to make a 'positive' copy, from which a hard wearing chromium plated 'negative' can be made. This copy is then used as a die in a heated press to make the plastic copies which we buy.

Replaying gramophone records

The earliest method of replaying a record was to cause a sharp steel needle to follow the groove, and the vibrations so produced would make a column of air vibrate to give a reasonable copy of the original sound. Surprisingly large volumes of sound would come from these all-mechanical gramophones playing 78 rpm records.

The modern microgroove record requires a needle or **stylus** of radius about 0·025 mm for monophonic, and 0·0125 mm for stereophonic recordings. An elliptical stylus tip follows the record groove better than a spherical one, but it is much more expensive to manufacture. The stylus tip may be made of sapphire, which has a useful life of about 30 hours, or of industrial diamond, which lasts about 1000 hours before wear is serious enough to damage the groove.

The stylus

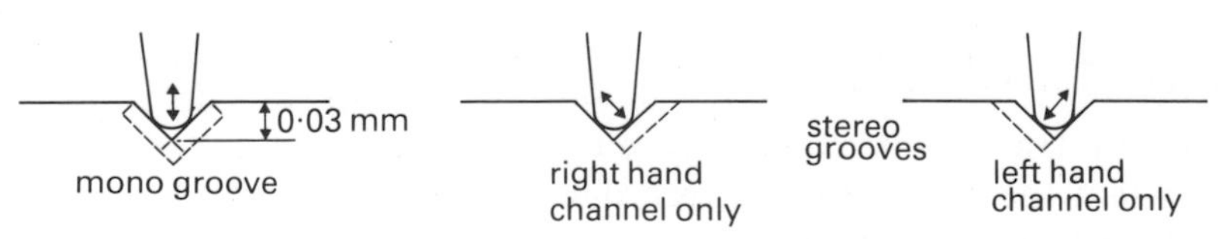

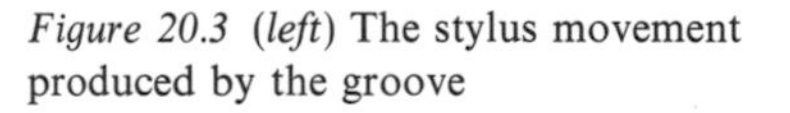
Figure 20.3 (*left*) The stylus movement produced by the groove

The stylus movement is shown in *figure 20.3*. This movement is converted into an electrical signal by the **cartridge**, which may be constructed in a variety of ways:

a high output, low cost **crystal**.
b lower output, better quality **ceramic**.
c very low output, high quality and expensive **magnetic**.

Both **a** and **b** make use of the piezo-electric effect (see unit seven), and have the basic construction shown, in its mono form, in *figure 20.4*. A simplified type of stereo magnetic cartridge is shown in *figure 20.5*.

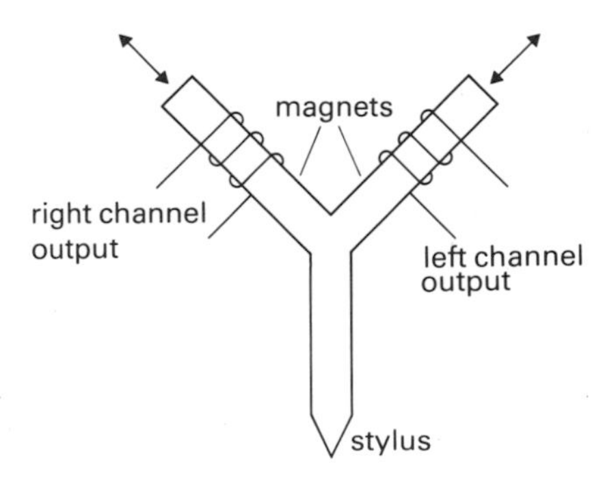

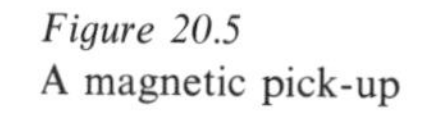
Figure 20.5
A magnetic pick-up

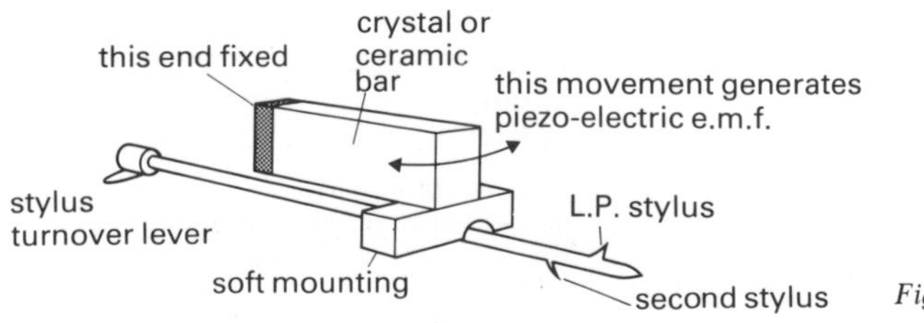

Figure 20.4 A crystal pick-up

The pick-up

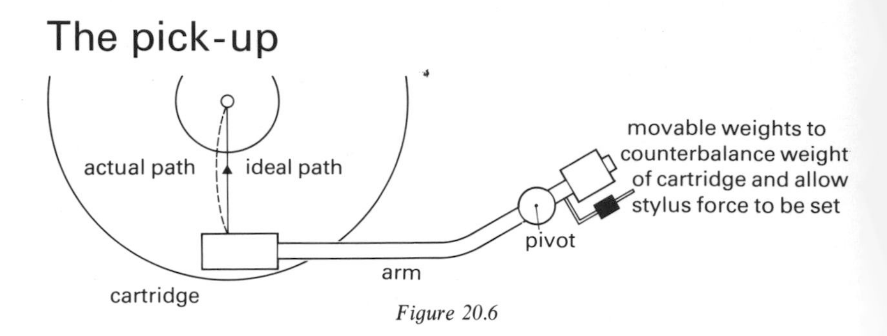

Figure 20.6

The modern pick-up arm (see *figure 20.6*) in a high quality record player is the product of years of skilled research and precision engineering. As the stylus is pulled across the record by the side of the groove, anything which hinders this movement will increase the wear of the record. This will spoil the ability of the stylus to follow the groove accurately. The pick-up arm must, therefore, be light and move in almost frictionless bearings. Also, by slightly bending the arm, the stylus will move across the record along approximately the same path taken by the cutter during recording. The vertical force needed to keep the stylus in the groove will depend upon the type of cartridge and pick-up arm. This force could be as low as 1 g for a top quality magnetic cartridge or up to 5 g for a ceramic type. The force can be set by means of movable weights, or sometimes springs, acting on the other end of the arm.

It is also found that the stylus does not sit in the centre of the groove, but exerts a slightly greater force on one wall. This force can be neutralised by applying a small sideways force to the arm. This is called the **anti-skating** or **bias compensating** force.

The turntable

As in the tape recorder, the replay of a record must be at exactly the same constant speed used for the recording. This constant speed is easily obtained by using a heavy turntable driven by an induction motor. *Figure 20.7* shows one method of driving the turntable from the motor at the different speeds required (usually 33 rpm and 45 rpm).

One problem which arises, is the low frequency mechanical noise known as **rumble**. This comes from the bearing on which the turntable rotates, and the noise is transmitted to the pickup. Motor vibration can also produce unwanted sounds and careful design of the motor mounting is necessary. The idler wheel can also transmit motor noise to the pickup and many manufacturers are now offering belt drive between the motor and turntable.

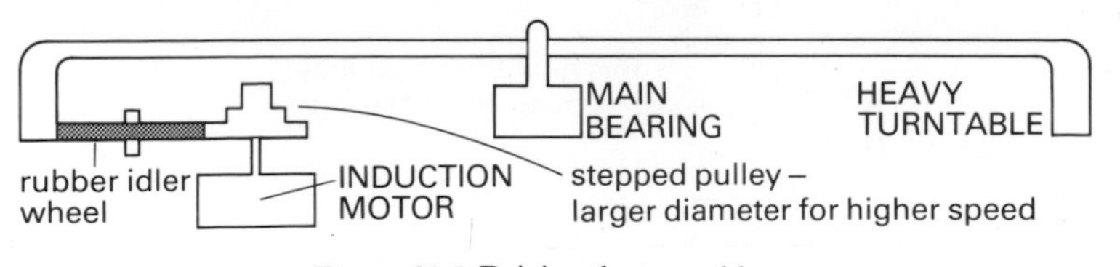

Figure 20.7 Driving the turntable

The complete audio system

A complete audio system would consist of the following units:

a record player–turntable and pickup.
b tape deck.
c radio tuner.
d pre-amplifier.
e power amplifier.
d loudspeakers.

All of these can be combined into one cabinet (apart from the speakers) and the manufacturers appear to want this to be called a 'music centre'. *Figure 20.8* gives a block diagram of the connections between each unit. If the units are obtained separately then the pre-amplifier and power amplifier would generally be in the same box. The purpose of the pre-amplifier is to enable the radio, turntable or tape deck to be switched separately through the amplifier, after the necessary frequency corrections and amplifications have been made. A magnetic cartridge delivers about 2 mV, whereas a radio tuner and tape deck produce about 250 mV, so they require less pre-amplification than the cartridge.

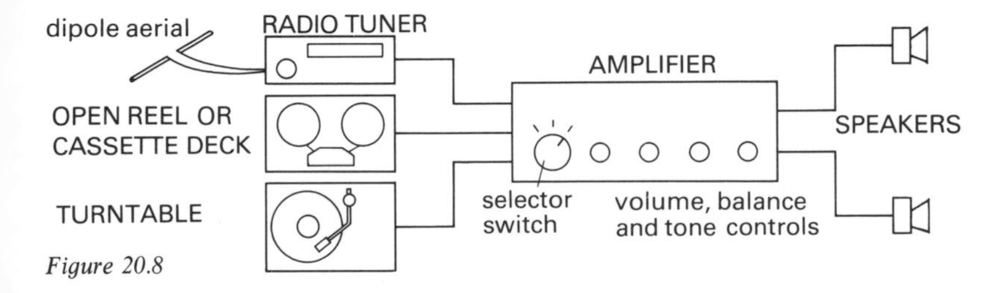

Figure 20.8

When connecting these units together, it is necessary to avoid what is known as an **earth loop**, as this will usually give rise to a low frequency hum in the loudspeakers. To avoid it, each of the four units should be earthed, not at their mains plugs but at a single point on the amplifier.

Another popular combination of units is the **tuner-amplifier**; this takes up much less space than a separate radio tuner and amplifier, because they can share the same power supply.

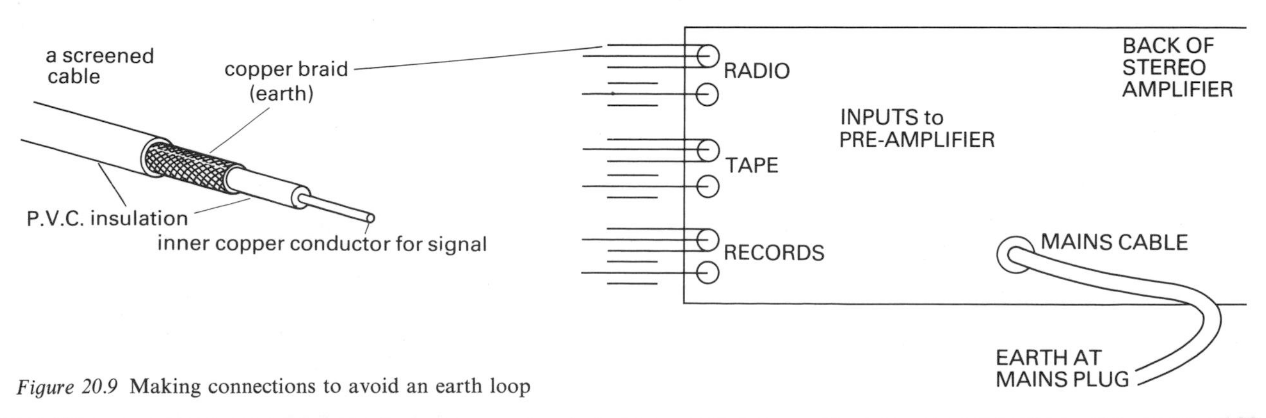

Figure 20.9 Making connections to avoid an earth loop

Unit 21

Integrated circuits

The electronics covered so far in this book has been dealing mostly with the operation of single or **discrete** components. In order to perform complex operations, such as in a calculator, it would be necessary to have hundreds of single transistors, capacitors and resistors. The construction of such a device from discrete components would be expensive and space consuming, and the chances of making a mistake would be fairly high. We know that the modern low priced calculator has only been possible because most of its components have been squeezed into a tiny slice of semiconducting silicon.

The silicon chip

Integrated circuits consist of a thin slice of silicon, or a silicon **chip** upon which has been etched a complex circuit of semiconductors, resistors and capacitors. For all except the very complex integrated circuits, the chip measures about 2 mm square by about $\frac{1}{4}$ mm thick. Most of what we see on the outside is packaging designed to give a device which is large enough to handle easily. Although there may be a number of rejects during the manufacturing process, once an integrated circuit works, it will be extremely reliable. One important effect of using integrated circuits is to make repair rather a different process; if anything goes wrong with a silicon chip, repair is impossible, so it must be replaced. The photograph shows a typical integrated audio amplifier with its cooling fins, which prevent an excessive rise in the temperature of the chip. A good example of a complex integrated circuit is the **microprocessor** or **microcircuit.** The microcircuit does nothing that could not have been done twenty years ago, but the reduction in size, cost, speed of operation and power consumption now means that the tasks can be carried out quite cheaply. Examples of microcircuit applications include:

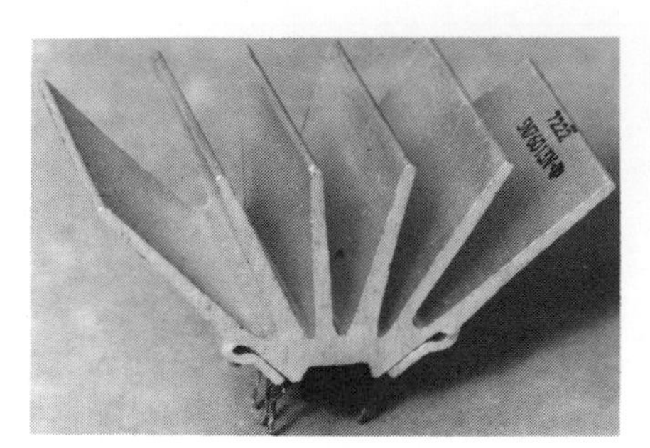

An integrated audio amplifier with cooling fins

a The replacement of assembly line workers by **robots**, which could carry out unpleasant, repetitious tasks such as paint spraying and spot welding.

b The use of a **word processor** can provide the typist with a stock of standard paragraphs from which a standard letter may be quickly assembled.

c The development of **electronic information services** such as the Post Office's Prestel system which works via the telephone, and Ceefax and Oracle which work via the television set. The range of information which could become available in this way is almost without limit. It might include an encyclopedia, advertisements for houses, jobs and goods, or information about the entertainment available in an area.

d The development of **driverless transport**, such as exists on London's Victoria Line underground railway.

e Computer assisted learning. This may take the form of recorded lesson packages with which students can interact by pressing appropriate reply keys on their computer keyboards. Alternatively, the Prestel system may eventually provide instruction via the telephone line, and again the students could make their own progress through the lesson by means of the keyboard.

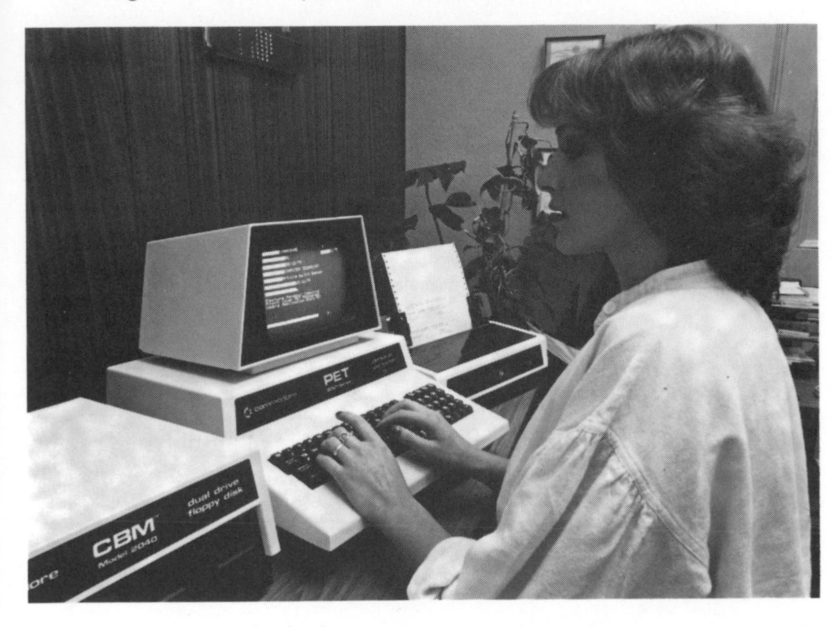

Left: A typical desk-top mini computer capable of performing a variety of tasks which until the advent of micro circuits would have required a computer of considerable size.

Electronic systems

The input signal

Information can be fed into an electronic system in many different ways, for example:

switched d.c. mixed frequency a.c. a.c. sine wave
modulated carrier pulses

If graphs are drawn to show how each of these varies with time, the results might look like one of the diagrams in *figure 21.2* (on the next page). See if you can match the graph with the correct description.

In this book, use has been made of the following signal sources:

audio frequency generator aerial microphone tape head pick-up multivibrator

Which of the graphs on the next page would best describe the output of each of these?

Passive components

The word **passive** may be used to describe a component which alters a signal but does not amplify it. Examples of such passive components are:

resistors switches tuned circuits capacitors logic gates transformers inductors relays

The action of all these components is described in the first half of this book. An alternative definition of a passive system is one which requires no external power supply. This definition would remove two items from the above list (logic gates and relays).

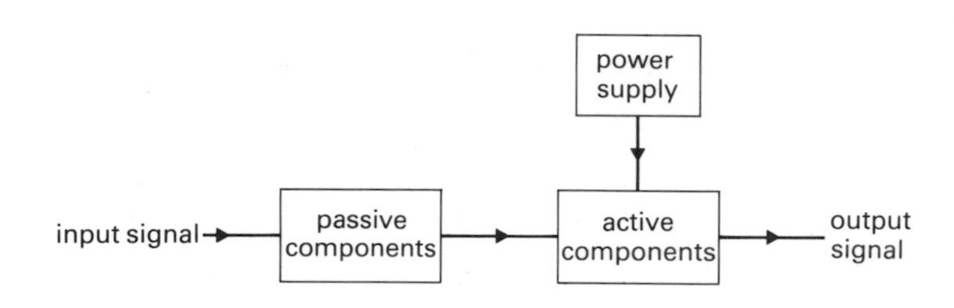

Figure 21.1 An electronic system

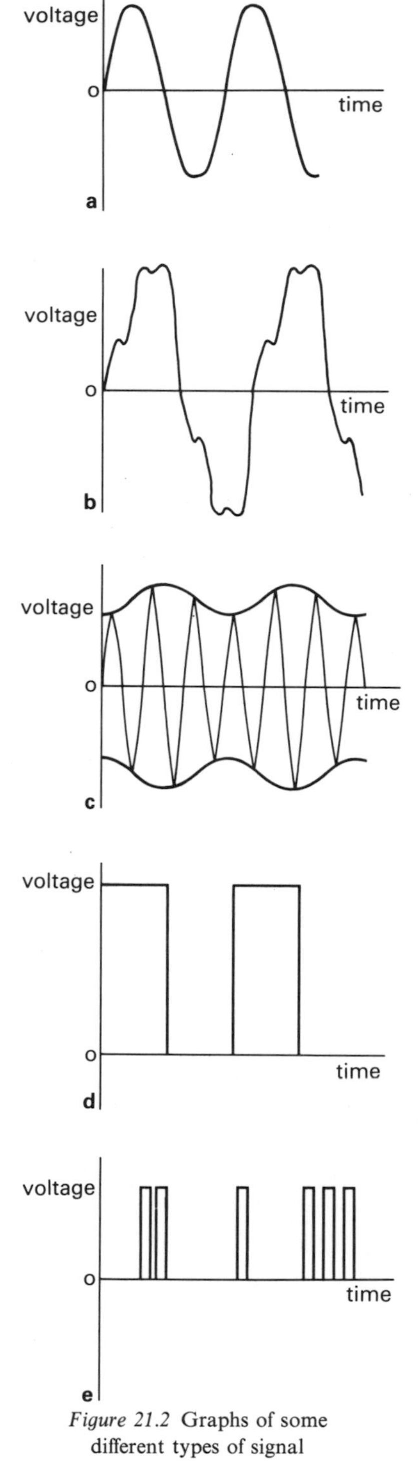

Figure 21.2 Graphs of some different types of signal

Active components

A component is said to be **active** if it is capable of amplifying a signal. Valves and transistors are the chief examples, although a diode may also be included under this heading.

Filters

A filter circuit may be active or passive, and it behaves in a way which depends upon the input signal frequency. A **high pass filter** allows only the high frequencies to pass through, stopping all the low frequencies. A **low pass filter** allows low frequencies through but stops all the high ones. A **band pass filter** is really a combination of high and low pass filters; it lets through a particular range of frequencies between an upper and a lower limit. Practical examples of these filters can be found in many record player amplifiers, where the high pass filter blocks the low frequency turntable rumble and a low pass filter removes high frequency record surface noise.

Power supply

Most active components need some sort of d.c. power supply, which may be one of the following:

dry battery rectified a.c. mains rechargeable battery solar cell

If the power supply is **unstabilised**, its voltage will vary with the operating current. A **stabilised** power supply will give a constant voltage, within limits. A Zener diode usually forms part of a stabilised power supply.

Output signal

The output signal from a system may be used to operate one of the following transducers:

headphones light-emitting diode electric typewriter loudspeaker
pen recorder television tube moving coil meter
ultrasonic transducer

Remember that in order to obtain the maximum possible transfer of energy from one system to another, the output impedance of one system must be equal to the input impedance of the other.

Block diagrams

Figure 21.3 shows the block diagram of a very common electronic system, an AM radio receiver. The names in each of these boxes should be familiar.

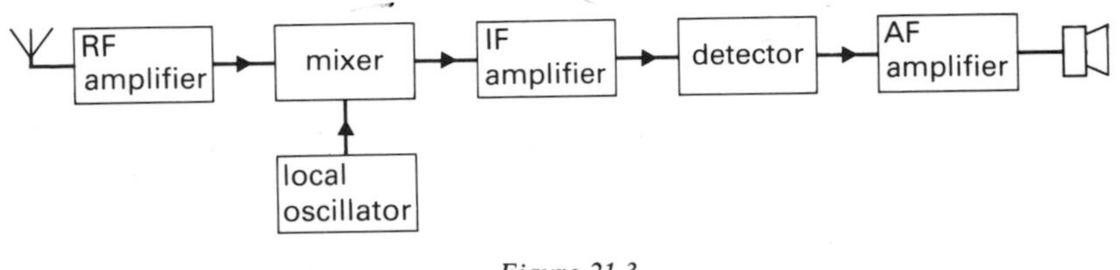

Figure 21.3

Figure 21.4 gives a simplified block diagram of a colour television receiver. The names in some of these boxes show that you have only just begun your study of the fascinating world of electronics!

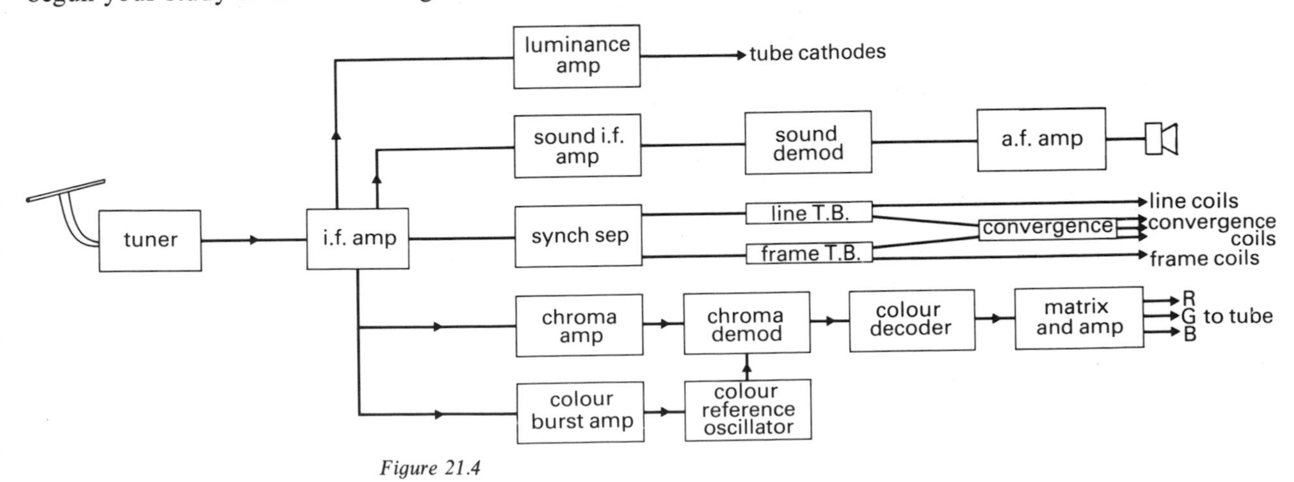

Figure 21.4

Test paper 1 (units one to six)

1 Name the three principal sub-atomic particles and give the charge on each one.

2 State the rule for electric charges.

3 What is the connection between charge and current?

4 What carries the current through **a** metals **b** solutions **c** gases?

5 Explain what is meant by conventional current.

6 If a current of 5 μA is to be written in the form $A \times 10^x$, what are A and x?

7 Write 0·01 V in millivolts.

8 Give three examples of non-linear resistors.

9 State the maximum power theorem.

10 **a** What is the value of a resistor with colour bands of red, violet, orange, silver?
b What is the colour code of 1 M2 5%?

11 Draw symbols for **a** polarised **b** non-polarised capacitors.

12 Give another word for a polarised capacitor.

13 State whether the following capacitors are polarised or non-polarised:
a tantalum.
b polystyrene.
c silvered mica.

14 Explain what is meant by a 'leaky' capacitor.

15 Draw a graph to show how the current increases when a p.d. is applied to an inductor.

16 Explain with the aid of a graph the meaning of the expression 'time constant'.

17 Show on the same graph how the reactance of an inductor and a capacitor vary with frequency.

18 What is a *choke* and why is it so called?

19 Explain the action of **a** a shunt **b** a multiplier when used with a moving coil meter.

20 What quantities would you expect to be able to measure with a universal testmeter?

21 Draw a graph, including suitable scales, to show two complete cycles of 50 Hz a.c. of peak e.m.f. 12 V.

22 What is the special property of **a** a perfect voltmeter **b** a perfect ammeter?

23 Describe the construction of a mains transformer core.

24 Explain **a** the efficiency of a transformer **b** turns ratio **c** impedance matching **d** autotransformer.

25 Where does a transformer 'lose' its energy?

26 Explain, with the aid of a diagram, what is meant by a rejector circuit.

27 **a** Where might you find an electron gun and **b** what is its function?

28 State the two functions of the anodes in a CRT.

29 What is the essential difference between a dual beam and a dual trace CRO?

30 **a** Explain the action of the X plates in a CRT and **b** give one reason why they are less sensitive than the Y plates.

Problems (units one to three)

1 Calculate the effective resistance of all possible combinations of three resistors having values 2 Ω 3 Ω and 4 Ω respectively.

2 Calculate the maximum safe current through a 25 Ω 1 W resistor.

3 A resistor is required to drop 50 V at 100 mA. Calculate its value to the nearest **a** E12 point **b** E24 point.

4 Write **a** 1000 pF in μF **b** 100 nF in pF **c** 4700 nF in μF.

5 Calculate the effective capacitance of
a 2 μF in series with 4 μF.
b 1 nF in parallel with 1000 pF.
c 500 pF and 1000 pF in series.

6 **a** Calculate the reactance of **i** 1 μF at 50 Hz **ii** 1 μF at 10 kHz.
b Calculate the value of capacitance which will pass a current of 300 mA when connected across 240 V a.c. mains.

7 **a** Calculate the value of R to be used with a 1 μF capacitor to give a 2 second time constant.
b If a 10 MΩ resistor is connected across a 4 μF capacitor charged to a potential of 100 V, how long will it take to discharge?

8 A 1000 μF capacitor is charged to a potential of 200 V. Calculate
a the charge stored **b** the energy stored.

9 Calculate the time taken for the current in a d.c. series circuit containing a 10 H inductor and a 100 Ω resistor to reach its final steady value.

10 Calculate the reactance of **a** 10 H at 50 Hz **b** 10 mH at 1 MHz.

Problems (units four to six)

1 A moving coil meter has a resistance of 1000 Ω and a F.S.D. current of 10 μA. Calculate:
a the multiplier value to give F.S.D. with 10 V.
b the shunt value to give F.S.D. with 3 A (2 significant figures).
c the meter sensitivity in ohms per volt.

2 A battery of e.m.f. 3 V and internal resistance 2 Ω is connected in series with resistors of 2 Ω and 5 Ω. Calculate:
a the terminal p.d. of the battery.
b the p.d. across the 5 Ω resistor.

3 A battery of e.m.f. 6 V and internal resistance 4 Ω is connected to resistors of 6 Ω and 3 Ω in parallel. Calculate the current through the 3 Ω resistor.

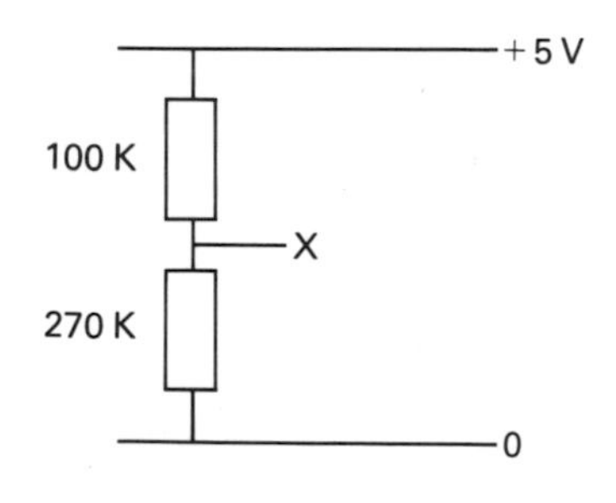

Figure A

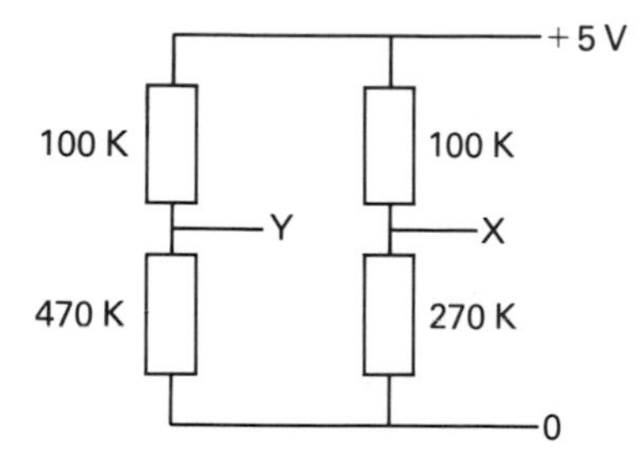

Figure B

4 Calculate in *figure A.*
a the potential at X relative to 0.
b the p.d. across the 100 kΩ resistor.

Calculate in *figure B.*
a potential at Y relative to 0.
b p.d. between X and Y.

5 Calculate the peak current and peak e.m.f. driven by 440 V r.m.s. mains through a 1000 Ω resistive load.

6 A transformer operating at 250 V on the primary delivers 40 V at 2 A to the secondary. Calculate the primary current if it has an efficiency of 80%.

7 An oscilloscope produces a vertical line 10 cm long when the Y amplifier is set to 0·5 V/cm. Calculate the r.m.s. value of the p.d.

8 If a time base sweep time of 2 ms draws $3\frac{1}{2}$ waves, calculate the frequency of the input.

9 Calculate the turns ratio on a transformer required to match 75 Ω to 600 Ω.

10 Calculate the maximum current which can be delivered to a 10 V secondary winding of 240 V primary transformer, if the maximum primary current is 10 A and the efficiency is 90%.

Test paper 2 (units seven to ten)

1 Explain the difference between primary and secondary cells and give one example of each.

2 Why can a nickel-cadmium cell be sealed, but not a lead-acid one?

3 Explain the *capacity* of a cell and give its units.

4 Upon what factors does a thermoelectric e.m.f. depend?

5 Name three piezo-electric materials.

6 Give two uses for a crystal oscillator.

7 Describe the difference between *photovoltaic* and *photoconducting* cells and give one example of each.

8 **a** Why is a quartz iodine lamp so called?
b Why must it be run at a much higher temperature than an ordinary filament lamp?

9 **a** Give two advantages of an LED over a small filament lamp.
b Give one advantage of an LED over a small neon lamp.

10 State whether the following properties apply to series or shunt wound motors:
a races off load.
b high starting torque.
c runs at almost constant speed.

11 Sketch the arrangement of coils and brushes in a compound wound motor.

12 **a** Which type of motor is used in mains operated record players?
b Give four advantages of this type of motor.

13 Sketch the output voltage for two revolutions of a 2 pole d.c. dynamo.

14 **a** Draw a simple diagram of a crystal microphone.
b Explain the need for a bimorph rather than a single crystal.

15 What is the approximate impedance of an **a** crystal **b** moving coil microphone?

16 Give an example of a microphone which can be used at the end of a long cable.

17 Draw the cross-section of a moving coil loudspeaker and label it.

18 **a** Explain the term *infinite baffle*.
b What is the drawback of the infinite baffle and how is it overcome?

19 If a loudspeaker is marked as having an impedance of 8 Ω, **a** explain what this means, **b** can an ohm-meter be used to measure it? Explain your answer.

20 Draw a circuit for a simple two way crossover.

21 Explain what is meant by two loudspeakers being *in phase.*

22 If a capacitor is being used to control current in an a.c. circuit, what two precautions should be observed concerning its type and rating?

23 Compare the advantages and disadvantages of the post office relay with the reed relay.

24 **a** Show how a variable resistor can be used as a potential divider.
b State two circumstances when the output voltage of the potential divider will not be proportional to the degree of rotation.

25 Explain how the back e.m.f. of a relay coil can be prevented from doing damage to a transistor.

Test paper 3 (units eleven to fourteen)

1 **a** Define *voltage gain* of a transistor
b Can it ever be less than 1?

2 What is meant by *matching* the load to an amplifier?

3 A student wishes to connect a 3 Ω loudspeaker to an amplifier designed to operate with a 15 Ω load. Comment on the following solutions he suggests:
a Use a 12 ohm resistor in series with the loudspeaker.
b Use a transformer of ratio 1:5.

4 Why is transformer coupling used between the stages of an RF amplifier but not usually between the stages of AF amplifiers?

5 How do you select a suitable capacitor for coupling two amplifier stages in terms of **a** its capacitance **b** its working voltage?

6 Explain the difference between positive and negative feedback.

7 Why do most amplifiers use some negative feedback?

8 When can positive feedback be **a** useful **b** a nuisance?

9 Name four semiconducting materials.

10 **a** What is meant by an *intrinsic* semiconductor?
b How does conduction in metals differ from conduction in semiconductors?

11 **a** What is a *hole*?
b How does a hole move?

12 **a** What is meant by an *n-type* semiconductor?
b Name the *majority carriers* in *n*-type silicon.

13 **a** Name the *minority carriers* in *p*-type germanium.
b What is the effect upon minority carriers of raising the temperature?

14 Why must a *p-n* junction start as a single crystal?

15 Draw a diagram to show the movement of charge carriers in a circuit containing a *p-n* junction on forward bias.

16 What is a varicap diode?

17 Draw a circuit which will produce full wave rectification.

18 A 300 mW Zener diode is marked 5V1.
a Explain what these numbers mean.
b Show how it can be used to give about 5 V from a 6 V battery.

19 What are *heat sinks* used for?

20 Give the three main uses for a transistor.

21 Draw a graph to show how collector voltage varies in an *n-p-n* transistor as the base voltage is increased from zero to 0·8 V.

22 Draw a graph to show how the collector current of an *n-p-n* transistor varies as the base current is increased from zero.

23 By means of a suitable graph show how distortion can be produced by overloading a transistor.

24 **a** Draw the circuit of a single stage transistor amplifier which will give similar results with different transistors of the *n-p-n* type.
b Explain the action of an emitter bypass capacitor in this amplifier.

25 When using an ohm-meter to test if a transistor has completely broken down, what readings might you expect to find?

26 Explain the terms **a** feedback **b** bandwidth **c** distortion, when applied to an amplifier.

27 **a** What are the properties of the ideal operational amplifier?
b How close to these ideals does a real operational amplifier come?

28 Explain the terms **a** inverting input **b** virtual earth and **c** closed loop gain, as applied to an operational amplifier.

29 **a** Sketch the common emitter output characteristic curves for a transistor.
b Draw a load line on the above curves and show how it is used to calculate the value of the load resistor.

30 Compare the action of the bipolar transistor with the unipolar FET.

Test paper 4 (units fifteen to twenty-one)

1 Draw a truth table for a two-input NAND gate.

2 Draw a truth table for a two-input exclusive OR gate.

3 Name the three gate symbols in *figure C*.

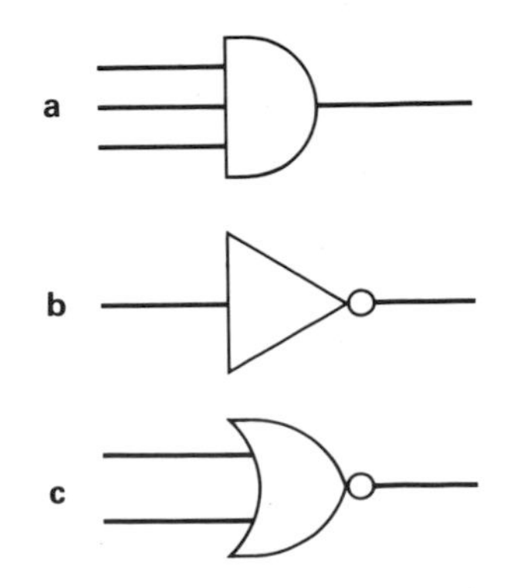

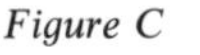
Figure C

4 Name the three types of multivibrator.

5 Which type of multivibrator can **a** divide by 2 **b** produce square waves **c** produce rectangular waves **d** produce a single square pulse?

6 **a** Write the decimal number 5 in binary form.
b Which decimal number is represented by the binary number 1001?

7 Sketch the graph of a damped oscillation

8 Draw a tuned circuit and write down its resonant frequency.

9 What sort of waves are produced by relaxation oscillators?

10 Draw square waves with mark-space ratios of **a** 1:2 **b** 2:1.

11 Why does sound have to modulate an RF carrier before it can be transmitted?

12 List the stages in the reception of AM waves.

13 How do we recognise that a radio receiver is not very selective?

14 **a** What is meant by a directional aerial?
b Give an example of a non-directional aerial.

15 **a** Draw the circuit of a simple diode receiver.
b State briefly what each component does to the signal.

16 Draw a simple diagram which shows the positions of the heads in a three head tape recorder. Label the heads and mark the direction the tape is moving.

17 **a** Can any sort of recording be made without recording bias?
b What happens if there is too much recording bias?

18 Explain the terms *wow* and *flutter*.

19 Why do tape heads need to be cleaned at regular intervals?

20 Draw a block diagram to show how a record groove can produce sound in a loudspeaker.

21 **a** List the operations performed by a monophonic pre-amplifier.
b What additional operations would you find on a stereo pre-amplifier?

Index

Solutions to problems

Units one to three

1 9 Ω; 0·923 Ω; 3·71 Ω, 4·33 Ω, 5·20 Ω; 2·22 Ω, 2·0 Ω, 1·56 Ω
2 0·2 A
3 470 Ω, 510 Ω
4 0·001 μF, 10^5 pF, 4·7 μF
5 1·33 μF, 2000 pF, 333 pF
6 **a** 3182 Ω, 15·9 Ω
b 3·98 μF
7 2 MΩ, 200 s
8 0·2 C, 20 J
9 0·5 s
10 3142 Ω, 62·8 kΩ

Units four to six

1 999 kΩ, 0·00333 Ω, 100 kΩ/V
2 2·33 V, 1·67 V
3 0·67 A
4 3·65 V, 1·35 V; 4·12 V, 0·47 V
5 0·622 A, 622 V
6 0·4 A
7 1·77 V
8 1750 Hz
9 1:2·83
10 216 A

Acknowledgements

Avo Limited p. 28, p. 29, p. 30
Baroness International Public Relations p. 141
Central Electricity Generating Board p. 34
E.M.I. Limited p. 136
Limrose Electronics Limited p. 106
Osram (GEC) Limited p. 23
C. Ridgers, V.R.U. p. 9, p. 10, p. 16, p. 17, p. 20, p. 23, p. 24, p. 25, p. 27, p. 47, p. 49, p. 104, p. 140